W9-CWV-730

The Evolution of California Manufacturing

• • •

Paul W. Rhode

2001

PUBLIC POLICY INSTITUTE OF CALIFORNIA

Library of Congress Cataloging-in-Publication Data
Rhode, Paul Webb.
The evolution of California manufacturing / Paul W. Rhode
p. cm.
Includes bibliographical references.
ISBN 1-58213-038-8
1. Manufacturing industries—California—History—20th century. I. Title.

HD9727.C4 R44 2001
338.4'767'09794—dc21 2001048377

Foreword

When PPIC opened its doors in 1994, a primary research objective was to understand California's emergence as a national and global economic power. A related objective was to understand the economic shifts that were under way during the mid-1990s, including the loss of high-wage jobs and the rapid increase in low-wage employment. To set the stage for detailed sector studies, PPIC commissioned economic historian Paul Rhode to take the long view of California manufacturing—from the early years of statehood in the mid-19th century up to the end of the 20th.

The product of that effort, *The Evolution of California Manufacturing,* offers a fresh portrait of the state's industrial history. Drawing on a unique and comprehensive database developed from U.S. *Census of Manufactures* files, the study reviews the state's emergence from a resource-based to a knowledge-based economy; its integration into the national economy; the effect of restrictive immigration laws on labor market patterns; the origins and development of high-wage sectors, such as aerospace and electronics; and the role of population growth in the economy's expansion and diversification. The study concludes with a surprising observation: There is little evidence to suggest that the 1990s was a period of especially rapid structural change. As the report makes clear, rapid change has been the norm for California's industrial development.

When combined with other PPIC work on the California economy—including Deborah Reed's *California's Rising Income Inequality: Causes and Concerns,* AnnaLee Saxenian's *Silicon Valley's New Immigrant Entrepreneurs,* Michael Dardia and Sherman Luk's *Rethinking the California Business Climate,* and Jon D. Haveman's *California's Vested Interest in U.S. Trade Liberalization Initiatives*—this study fleshes out a statistical portrait that was unavailable five years ago. As a result of this work, we know more about the state, its strengths and vulnerabilities, and how current changes fit into the larger picture of an evolving, diverse, and robust economy. As in the previous reports, the scholarship is sound, the interpretations are cautious, and the findings are as remarkable as the California economy itself.

David W. Lyon
President and CEO
Public Policy Institute of California

Summary

Over the course of the 20th century, California's manufacturing sector experienced a remarkable transformation. Between 1899 and 1997, this sector's real value added grew at an annual rate of over 4.8 percent, or 1.8 percent above the national rate. What had been an unimportant industrial location in 1900 became the nation's number one manufacturing state by 1977. By the end of the 20th century, California's manufacturing value added was more than 50 percent above its leading rival. During certain key periods such as 1939–1958, manufacturing served as the driving force for California' growth, but more generally its expansion was part of a balanced growth process in a state whose gross domestic product would rank now among the world's top eight national economies. A simple "march of time" explanation does not do full justice to the scope or timing of this transformation. Rather, a careful analysis of the historical evidence suggests that this transformation is the result of a complex interplay among regional, national, and international forces.

Long-Run Patterns and Processes

Although it is impossible to provide a simple, complete explanation of California's rise to national industrial leadership, several long-run processes have shaped the pace and pattern of California's manufacturing development.

Shift from Resource-Based to Knowledge-Based Growth

Early in its economic history, California's industrial growth depended on the state's distinctive natural resource endowment. This resource base fostered the development of canning, winemaking, lumber production, and petroleum refining. As late as 1935, resource processing industries accounted for over one-half the industrial activity in California. Knowledge-based industries took off after the mid-1930s, driven initially by the growth of the aircraft industry and then by electronics in the 1950s and 1960s. By 1997, high technology accounted for one-quarter of California's manufacturing production workers and almost half of value added.

Product Market Integration

As California's home markets grew following the Gold Rush, many new firms and industries served these markets profitably. As transportation costs fell, however, local manufacturers often found they could not compete with outside producers. At the same time, lower transportation costs allowed several signature industries—such as winemaking, canning, and aircraft—to pursue export markets. Over time, California's manufacturing sector became more integrated into the national and, later, global economy.

Factor Market Integration

The 1882 Immigration Restriction Act cut off labor inflows from Asia, forcing California to forge stronger labor connections with the American Midwest, which was characterized by high wages and high levels of human capital investment. As a consequence, California's manufacturing development was channeled along a high-wage path. More recently, increased labor inflows from Latin America and East Asia have encouraged the growth of a cluster of low-wage activities such as apparel and leather goods manufacturing.

Realizing Economies of Scale

During the late 19th and early 20th centuries, the small size of the region's economy limited local development of the emerging large-scale industries and constrained the scope of manufacturing activities. When California's population started to grow rapidly again after 1900, its economy began to realize internal and external economies of scale. Mass production industries such as automobiles and rubber tires entered the state to produce for the western market and, more generally, the scope of industrial activities widened. Realizing external economies of scale was especially important among the state's high-technology firms, such as its aircraft and electronics producers. Less than a decade after Donald Douglas established a local aircraft firm in 1920, Caltech emerged as a center for aeronautical research. By the mid-1930s, eastern firms were relocating to California to tap the local pools of engineering talent and venture capital. Eventually, the electronics industry benefited from the technological and demand spillovers from the aircraft industry. By realizing such economies of scale, California's population and its manufacturing sector expanded together in a mutually reinforcing process over the past century.

Most of the potential in the dynamic has been realized as the state's economy has matured. As a result, the current role of the manufacturing sector in the state's economic growth is not that of the "leading man" as it was in the 1940s, 1950s, and 1960s; today, it would be better characterized as a major supporting player or key member of an ensemble cast. Nonetheless, its continued success remains important if the California economy is to sustain its historic pattern of balanced growth and to avoid becoming overly dependent on a single sector such as services to drive its economic expansion.

Recent Trends in California Manufacturing

This historical analysis brings into sharper relief several noteworthy recent trends:

- Over the past 25 years, the state's manufacturing sector has displayed signs of increased polarization. The high-technology and labor-intensive sectors have expanded, but there has been little growth in the "middle."
- The long-run trend in California for salary growth to outpace wage growth has accelerated, creating widening earnings gaps.
- For the first time in the 20th century, industries relying on inflows of low-paid immigrants experienced sustained growth.
- The state's long-standing specialization in the transportation sector has disappeared. This appears largely a result of the contraction of the aerospace sector following the defense cutbacks and recession of the early 1990s.

- By many measures, California has lost its historical character as a high-wage, labor-scarce region. Its productivity and labor earnings have converged to the national levels.
- California manufacturers have become increasing focused on export markets. In 1997, direct exports accounted for 16 percent of the value of the state's manufacturing shipments, four times their share three decades earlier.
- California has maintained its tradition of fostering a vibrant small-plant culture. Notably, the prevalence of small enterprise, which was historically a cause of concern, has become as celebrated characteristic of the state's manufacturing sector.

Taking a long view reveals other important insights. The downturn of the 1990s captured the attention of California's economic and political observers, as the forces of deindustrialization, which had buffeted the national economy since the 1970s, began to more fully affect the state. The historical evidence indicates that, contrary to many contemporary claims, there is little to suggest that the past decade was a period of especially rapid structural change. Indeed, change has been the one true constant in California's remarkable industrial transformation.

Contents

Figures

Tables

Acknowledgments

I would like to acknowledge several people who contributed to this research project. I would first like to thank David Lyon for his generosity in providing PPIC's resources for this effort and for giving me the opportunity to visit the institute. I would also like to thank Michael Teitz for his advice, support, and patience, without which this project would not have been possible. I have also enjoyed many stimulating discussions with PPIC's researchers about California's past, present, and future. Peter Richardson's excellent editorial assistance helped turn a sprawling early draft into a tighter, much more readable monograph. The PPIC staff has been first-rate. I would also like to thank Robert E. Gallman and Arthur Woolf for providing some of the data used in this study.

All errors in this report are the sole responsibility of the author.

1. Introduction

Over the past 150 years, California has experienced a remarkable economic transformation. What had been a relatively unpopulated outpost in the late 1840s has emerged as an economic colossus. Today, the state makes up about one-eighth of the U.S. population and an even larger share of the nation's economic activity. It is, by far, the leading industrial state, and its gross domestic product would rank it among the world's top eight national economies. But a simple "march of time" explanation does not do full justice to the story of California's economic transformation. Even after the transcontinental telegraph and railroad were completed in the 1860s, the state's population share rose little, and its share of national income actually fell between 1880 and 1900 (see Figure 1.1). Indeed, some perceptive local observers portrayed California as slipping into a state of dormancy during this period (Wheeler, 1911,

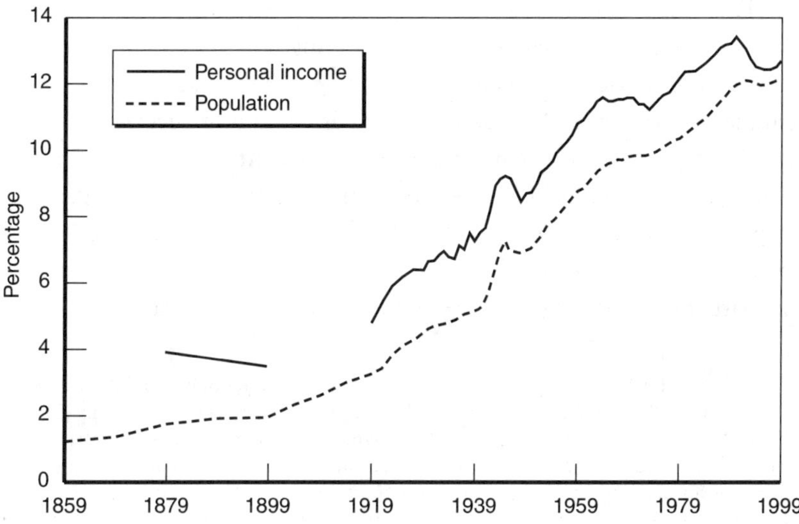

Figure 1.1—California's Share of National Personal Income and Population

pp. 167–168). California's emergence as a global center of economic activity was primarily a 20th-century phenomenon.

An analysis of the manufacturing sector helps us understand this phenomenon. Over the past two centuries, industrialization has been the driving force in most countries experiencing "modern economic growth."[1] As a result, policymakers have often valued manufacturing more highly than other economic activities. And they have taken steps to promote the development of this "strategic" sector, often at the expense of other production activities, consumers, taxpayers, and the environment.

To date, there has been little systematic analysis of California's long-run industrial development. One reason is that government and private statistical agencies cannot afford to devote their scarce resources to conducting studies with a long view. A second reason is that economic historians, who have long been preoccupied with issues of 19th-century growth and development, are only now turning their attention to the spectacular economic expansion of the American West over the 20th century. A third reason is that the data for such an investigation have not been available in a form suitable for ready analysis.

This study fills an important gap in our current understanding of California's economic development by offering policymakers and researchers a primer on the long-run evolution of the state's manufacturing sector. Its first and most basic task is to "get the facts straight" by constructing comprehensive, consistent data series on the level, structure, and rate of growth of manufacturing employment in California. These new series allow for fuller analysis of the patterns of growth and structural change of the California manufacturing sector and for sharper comparisons with the other regions and the nation as a whole.

---

[1]That is, in most nations experiencing high, sustained rates of per capita income increase since 1800, the manufacturing sector has grown faster than the economy as a whole. The sector's shares of output and employment have greatly expanded, often making manufacturing the largest sector in the economy. Typically, manufacturing has accounted for between one-fifth to one-third of total employment and output. Manufactured products have usually represented an even larger fraction of goods entering interregional or international trade and hence have served as key sources of export earnings. Over this epoch, manufacturing activities have often paid labor and other factors of production higher returns than the agricultural and service sectors.

This study has the following form. Chapter 2 introduces the new dataset on manufacturing activity in California and the nation as a whole, discussing the problems confronted in its construction and the advantages it offers over the previously available series. Chapter 3 uses the new data to chart the size and growth of the California manufacturing sector over the past 150 years. Chapter 4 analyzes the major structural transformations in the state's manufacturing sector and Chapter 5 explores the sector's changing comparative advantages. Chapter 6 highlights California's role as a pace-setter for many of the major trends in national industrial development over the 20th century. Chapter 7 explores the relatively higher earnings and productivity of labor in California's manufacturing sector and Chapter 8 investigates the greater prevalence of smaller-scale plants in the state. Chapter 9 provides a qualitative analysis of the forces shaping California's industrial experience. Chapter 10 explores the evolving role of the manufacturing sector in the California economy and Chapter 11 concludes by summarizing the main findings of this historical study.

2. Data and Methods

Our understanding of California manufacturing and its evolution has been handicapped by severe deficiencies in the historical record. The U.S. Census Bureau did not provide reasonably consistent data on California manufacturing at the aggregate sectoral (two-digit) level until 1947, making the analysis of the state's manufacturing development over much of the American period extremely problematic. In addition, shifts in Census categories make it difficult to form a clear picture of the state's evolving industrial structure. In place of analysis based on concrete evidence, oversimplified generalizations—such as the statement that the state's economy was based entirely on agriculture before the advent of the aircraft industry—hold sway.

To overcome these data difficulties, I constructed a comprehensive dataset on industrial activity in the United States and California for the period since 1859. The information on 375 ± 75 industries defined at the four-digit level was drawn from the *Census of Manufactures* and assembled into consistent paneled time series. The years covered in this study include 1859, 1869, 1879, 1889, 1899, 1904, 1909, 1914, 1919, 1921, 1923, 1925, 1927, 1929, 1931, 1933, 1935, 1937, 1939, 1947, 1954, 1958, 1963, 1967, 1972, 1977, 1982, 1987, 1992, and 1997. Among the variables included are the number of establishments, production workers, wages, nonproduction workers, salaries, cost of materials, value of production, value added, capital, and horsepower.[1]

These data have several advantages over the previously available series. They provide consistent, relatively disaggregated information over a long historical sweep. The state-level data are linked to national data to

[1] In this study, I have used virtually the entire collection of the *Census of Manufactures* and *Annual Survey of Manufactures* and will take the liberty of not citing exact volumes and page numbers unless referring to a specific quote or special publication. The most current official data are available at http://www.census.gov/epcd/www/econ97.html.

permit direct comparisons of the structure and characteristics of California's industrial sector with those of the United States. In addition, the four-digit industrial data employ a consistent set of industrial classification (SIC) codes and can be aggregated into the standard two-digit and three-digit categories with relative ease. Heretofore, there were no consistent machine-readable data about manufacturing activity for the nation as a whole, much less for California. Another advantage is that, where possible, the cross-sectional data are paneled (that is, linked over time) to allow for more explicit comparisons of growth performance at a disaggregated level. Finally, these data present a much clearer picture of the state's industrial structure in 1899 and 1939, which many contemporary observers and later researchers considered key turning points in California's economic development (Nash, 1972).

In much of the analysis that follows, the preferred measure of the level of activity is the number of wage-earners or production workers. This attention is justified on several counts. First, this number reflects the real economic activity of actual human beings, unlike value added, which the Census officials calculate as a residual of the estimated value of production minus the estimated cost of materials and which includes price variables. Second, work on the production floor is arguably closer to the essence of manufacturing. (Note that I am not arguing that nonproduction work is "unproductive" labor or any less valuable than production work.) Manufacturing is formally defined as involving "mechanical or chemical transformation" or the "assembly of components" to create new products, physical activities inherently requiring some "production" workers. Much of what "nonproduction" manufacturing employees do—for example, distributing manufactured products or keeping books—could easily by categorized as a service activity. Third and most fundamentally, data on production workers are available on a reasonably consistent basis for the greatest number of industries over the longest time period. Data on total employment (including both production and nonproduction workers) are not consistently available at the detailed level in the main state tables until 1939.

Disclosure Problems

The data published in the Census were subject to certain nondisclosure rules. The Census Bureau did not reveal detailed information on the activities in a given area of a four-digit industry after 1890 if there were fewer than four firms in operation and after 1947 if there were fewer than 150 employees. Such rules seldom led to problems with the national data, but they do create headaches when studying California, especially if one is interested in activities involving a small number of plants. Fortunately, the Census almost always reported the number of establishments in the state in each industry. In addition, Census officials did not keep their secrets very well in the age before computers, making it frequently possible to find information in other parts of the Census that reveal exact employment or value added for industries left "undisclosed" in the California table.

Where this was not possible, I estimated employment or value added in the individual "undisclosed" industries using a procedure that employs the most specific information available and has a zero mean error across the "undisclosed" industries in each Census year. The Census provided information of the aggregate level of activity in all undisclosed industries in California in each year (which can serve as a control total). It also reported, for each undisclosed industry, the number of establishments in the state and the total level of employment and output in the residual "all other states" category. The inference procedure essentially first makes the California establishments in each "undisclosed" industry as large as those in the residual category and then inflates or deflates the level so that the sum over all "undisclosed" industries in the state in a given year matches the California control total.

In the typical year between 1889 and 1939, inference techniques were required for about 22 percent of California industries, accounting for 6 percent of production workers and 7 percent of value added. Between 1947 and 1963, the Census withheld detailed information for a greater share of four-digit industries (roughly 40 percent, accounting for 23 percent of production workers and 22 percent of value added) but typically published the output and employment data at the three-digit level. Although the samples for these years required more inference, it could be done with far great accuracy. I am reasonably confident that the resulting picture of California's industrial structure is close to a true representation.

Coverage Changes

The coverage of the published Census was quite inconsistent before 1935. In the 19th century, the U.S. Census Office enumerated numerous hand and neighborhood trades, custom operations, and, at times, mining activities in the *Census of Manufactures.* In the early 20th century, the Census included large business activities such as automobile repair shops, railroad car repair performed by railroad companies, illuminating gas, motion picture production, and other activities that were not considered "manufacturing" by 1939. For a brief time in the 1920s, the Census even counted motion picture production as manufacturing. The practice stopped after 1929, leading to an exaggerated picture of the downsizing of California manufacturing during the Great Contraction. With an eye to creating consistent long-run series, I typically excluded such activities from the analysis.

One problem deserves special mention. After 1899, the Census excluded a wide range of smaller-scale activities (custom and hand trades) from manufacturing. In 1899, these activities constituted about 60 percent of national establishments, 12 percent of the national employment of production workers, and 14 percent of value added.[2] It was easy to drop the hand and neighborhood shops, which were previously enumerated as separate industries. Handling the exclusion of custom operations within ongoing industrial categories was much more difficult. As a result, I typically present two sets of statistics for 1899: one consistent with the 19th-century data that include custom operations and another set consistent with the 20th-century data that exclude them.

Category Shifts

The Census began employing the modern SIC system in 1947 and started shifting to the North American Industrial Classification System

---

[2]U.S. Bureau of the Census, *Historical Statistics of the United States* (1975).

(NAICS) in 1997. The pre-1947 system classifications were broadly similar to the modern SIC categories but there were sufficient differences, particularly in the treatment of metals and machinery, to create taxonomy problems. In addition to these large changes, the census-takers continuously made small adjustments to their definitions. This made paneling the data especially difficult and time-consuming. In performing this task, I generally chose to retain consistency in the disaggregated series at the cost of suffering breaks over time. The payoff from these efforts is a much clearer picture of the long-run evolution of California's manufacturing sector.

Inclusion of the 1997 data represented another significant challenge and an additional important contribution of this study to an understanding of the contemporary California economy. Reflecting the current age of globalization and rise of the new economy, the 1997 *Census of Manufactures* began using NAICS in place of the tried-and-true SIC system. The NAICS is designed to make industrial statistics comparable within the North America free trade zone (Canada, Mexico, and the United States). It makes a number of category changes to update the industrial structure, for example, shifting printing and publishing out of manufacturing into a new information sector. To provide a bridge to earlier series, the Census Bureau released detailed national conversion tables between the NAICS and SIC industries in June 2000. It also produced state-level statistics using the two-digit SIC categories for the number of establishments, employees, payrolls, and value of shipments. But the Census Bureau did not provide state-level two-digit statistics for two variables of great interest for the study of industrial development—the number of production workers and manufacturing value added. This study fills this crucial gap by laboriously deriving estimates for California production workers and value added in 1997 at the two-digit SIC level.[3] This effort provides a clearer picture of the evolution of the state's industrial structure over the past 150 years.

---

[3]To provide 1997 California data under the SIC system, I used national bridge tables allocating the number of establishments, employment, production workers, and value added of the six-digit manufacturing activities under the NAICS system across the four-digit SIC sectors.

One final issue: Unless otherwise noted, I converted all nominal data, such as value added and payrolls, into "real" values using the gross domestic product deflator, which reflects the prices of currently produced goods and services at the national level. The deflator is based on the official Department of Commerce series back to 1929 and the Balke-Gordon series from 1929 back to 1869 (Balke and Gordon, 1989). The 1859 prices were derived by linking these series with the price deflator implicit in the commodity production series reported in *Historical Statistics of the United States* (U.S. Bureau of the Census, 1975). Using a national deflator does not adjust for price differences between California and the country as a whole, making certain productivity comparisons less reliable.

3. The Growth of California Manufacturing

A century ago, California was unimportant as an industrial area. In 1899, it ranked 13th among the U.S. states in terms of manufacturing value added and 14th in terms of number of production workers. Today, California is the nation's leading industrial state, with manufacturing value added equal to roughly 11 percent of the national total. That share is half again as high as that of its nearest rival, Texas. California's dramatic rise is charted in Table 3.1, which documents the number of establishments, total employment, total payrolls, the number of production workers (wage-earners before 1947), total wages, and the value added of the state's manufacturing sector.

California's performance during the 20th century would have been difficult to predict based on the region's lackluster industrialization experience over the late 19th century. Although manufacturing growth maintained a rapid pace nationally, it was slowing in the newly settled western state. Indeed, during the 1890s, the growth rates of both number of production workers and real value added in California lagged well behind the nation as a whole. As the century turned, however, growth in the state accelerated. Between 1899 and 1904, the number of manufacturing establishments in the state rose by 35 percent and the number of production workers increased by 24 percent. By way of contrast, the number of establishments rose by only 4 percent nationally and the number of production workers by 15 percent. Another sign of the boom at the turn of the century was that California's relative wages in manufacturing jumped up sharply, reversing a long-run tendency to converge toward the national average over the late 19th century. The state's wage premium relative to that of the nation rose from 19 percent in 1899 to 34 percent in 1904.

Table 3.1

California Manufacturing Activity, 1859–1997

Year	No. of Estab-lishments	No. Employed	Payrolls	No. of Production Workers	Hours Worked	Wages	Value Added
1859	1,218	—	—	6,052	—	5,047	10,792
1869	2,763	—	—	21,890	—	10,727	26,457
1879	4,231	—	—	39,525	—	18,427	38,510
1889	4,695	65,828	38,444	58,286	—	31,049	74,657
1899	6,443	78,995	43,301	71,976	—	35,954	74,328
1899'	4,925	78,112	42,721	71,559	—	35,867	86,940
1904	6,755	101,871	70,868	90,404	—	57,267	114,739
1909	7,522	127,348	94,319	102,386	—	72,664	187,173
1914	9,446	155,906	123,951	121,983	—	90,918	240,382
1919	10,155	265,275	338,757	217,312	—	268,033	705,859
1921	8,502	213,629	317,381	177,398	—	242,082	591,175
1923	9,047	267,135	405,753	220,260	—	309,357	849,444
1925	9,433	—	—	227,567	—	315,425	880,320
1927	9,863	—	—	237,520	—	335,082	964,627
1929	11,839	317,469	503,364	264,418	—	375,749	1,198,079
1931	9,956	—	—	192,970	—	250,692	772,279
1933	8,334	—	—	181,138	—	180,149	594,620
1935	10,345	284,096	358,120	239,101	—	265,645	808,130
1937	10,861	358,083	503,735	302,189	—	389,132	1,091,597
1939	11,558	357,098	533,744	271,290	—	358,734	1,122,545
1947	17,648	663,872	2,064,523	530,283	1,070,270	1,519,255	3,994,981
1954	24,509	1,053,255	4,807,399	773,686	1,534,909	3,151,410	8,597,453
1958	28,735	1,217,300	6,876,300	838,671	1,656,700	4,107,200	12,048,000
1963	32,201	1,397,600	9,612,200	897,500	1,791,400	5,195,200	17,185,000
1967	31,962	1,583,500	12,514,500	1,044,900	2,089,700	6,877,800	23,393,600
1972	35,699	1,545,100	15,483,100	1,020,000	1,974,900	8,430,400	31,175,200
1977	45,289	1,751,500	24,671,500	1,142,600	2,224,200	13,150,500	54,862,400
1982	47,625	2,004,800	42,636,400	1,209,400	2,317,900	20,564,800	94,374,000
1987	49,935	2,103,400	57,133,600	1,276,200	2,432,500	25,694,100	132,403,500
1992	50,478	1,946,700	65,243,700	1,114,900	2,248,700	26,862,500	156,937,400
1997	49,079	1,870,016	—	1,193,550	—	—	204,119,356

NOTE: 1997 number of establishments and number employed exclude central administrative offices. Payroll, wages, and value added are in thousands of current dollars. 1899 and earlier figures include custom operations; 1899' and later figures exclude them.

Over the next four decades, a series of dramatic events shocked the state's manufacturing sector, making short-run growth comparisons precarious. The 1906 San Francisco earthquake led to the destruction or relocation of hundreds of manufacturing establishments in the Bay Area and caused employment statewide to shift temporarily into construction-related activities. Over the 1914–1919 period, military demands stimulated a 14-fold expansion of employment in the state's shipbuilding industry. After the end of World War I, the shipbuilding industry rapidly downsized, shedding 80 percent of its employees. Fortunately, other parts of the state's manufacturing sector picked up the slack, leading aggregate activity to surpass its 1919 peak by 1923. Growth was especially brisk in the late 1920s. But after 1929, the worldwide Great Depression subjected the state's economy to another enormous shock. Between 1929 and 1933, employment of production workers contracted by 31 percent in the state. The downturn was about equally severe in California as in the nation as a whole, but the state's recovery over the 1933–1939 period was considerably more robust.

These economic shocks suggest that we should to take a longer view of California's manufacturing growth in the early 20th century. If the period from 1899 to 1939 is considered as a whole, the number of manufacturing establishments in the state grew almost 2.3 times, manufacturing employment 3.8 times, and real value added 6.7 times. Even with the sharp decline in shipbuilding activity following World War I, employment in California manufacturing expanded both in the 1920s and the 1930s, in contrast to the national contractions over these periods.

Between 1939 and 1947, employment in the state's manufacturing sector almost doubled from 271,000 production workers to 530,000. The growth of California manufacturing during World War II appears especially dramatic against the backdrop of the Great Depression. Many observers treat this war as the watershed in the state's development. But it is crucial to recognize that the state's rapid industrial expansion began well before 1939. Indeed, if the fast growth of the 1899–1929 period had continued, employment and output would have remained above their actual levels even at the wartime peak. Note that growth between 1947 and 1958 was also impressive. Although the percentage changes are not as great as those between 1939 and 1947, the absolute magnitude of the increases of real value added and production workers exceeded those of the period including the war.

Between 1958 and 1972, manufacturing growth in California slowed relative to that of the country as a whole. In the late 1950s, the state's share of national manufacturing employment and output reached a plateau that lasted through the mid-1970s. Note that the slowdown in the state's industrial growth was followed, with a short lag, by a slowing of its relative income and population growth (see Figure 1.1). This pattern indicates that manufacturing activity and total income grew more rapidly in other regions of the country than in California during the latter part of its "Golden Age," or the Brown-Reagan years.

After 1972, California's share of national manufacturing activity began to rise again. Between 1972 and 1987, the number of manufacturing establishments in the state increased about 42 percent compared to 12 percent nationally. California's increasing shares of U.S. industrial activity were the product of an acceleration in growth locally and, more important, of the "deindustrialization" process occurring outside the state. The disparity between the performances of the California and U.S. manufacturing sectors was especially great between 1977 and 1982. During this period, the total number of production workers increased by 5.8 percent in the state but declined by 9.5 percent nationally.

California's relative success was due to a boom in a handful of sectors headlined by electrical components, computing equipment, printing, and scientific instruments—in other words, the information technology sector. A broad swath of the state's manufacturing sector did not enjoy this prosperity. Instead, it was mired in the same problems—recession and rising international competition—that were adversely affecting manufacturing across the country. Industrial restructuring during the 1980s led virtually all of California's integrated steel, auto assembly, and rubber tires plants to close their doors. The stagnating and declining industries in the state's manufacturing sector represented a major share of total activity. In 1977, those industries had employed over half of California's production workers.

With the severe recession of 1990–1994, the process of deindustrialization caught up with the state's manufacturing sector. Aerospace (aircraft, guided missiles, and search and navigation equipment), of course, were especially hard hit. But the declines were more broadly based than just in the sectors affected by post–Cold War cutbacks in military demands. Although California manufacturing had experienced robust recoveries after even more severe reductions in military demands following World Wars I and II, no such recovery occurred in the 1990s. In 1997, the number of production workers in California manufacturing remained below the peak level (1.28 million) achieved in the late 1980s. Despite these problems, California accounted for 10.7 percent of national manufacturing value added and 9.7 percent of production workers in 1997. The state's hard-won mantle as the nation's leading industrial state appears secure for now.

Although California has become the leading industrial state in absolute size, manufacturing has always played a relatively smaller part in its economy than in the nation as a whole. Indeed, in every *Census of Manufactures*, California's share of the national number of production workers is below its share of the national population. And the state's share of national manufacturing value added is less than the state's share of national income. Table 3.2 charts movements in the number of production workers per 1,000 people in California and the United States as a whole from 1859 to 1997. The table also displays (a) the ratio of California's share of the national production workers to its share of the national population, which, as a shorthand, will be called the relative employment ratio, and (b) the ratio of California's share of national value added to its share of personal income, which will be called the relative output ratio.[1] Note that if the employment ratio exceeds the output

[1]Data on state personal incomes first became available for California in 1880. It would be possible to calculate the ratio of manufacturing value added to personal income in the state and the nation, but this could lead to ill-advised inferences about the share of manufacturing in the overall economy. In income accounting concepts, manufacturing value added is a building block in the calculation of gross product, which is considerably larger than personal income. The relevant components of personal income, namely, earnings from manufacturing, are typically on the order of two-thirds of the size of manufacturing value added. Unfortunately, official series on personal income by state (including one-digit earnings) are available only after 1929 and do not provide earnings

Table 3.2

Measures of Prevalence of Manufacturing in California, 1859–1997

	Production Worker/ 1,000 Population		Ratio of California/United States	
Year	California	United States	Production Worker/ Population	Value Added/ Income
1859	15.9	36.8	43.3	—
1869	39.1	45.9	85.2	—
1879	45.7	51.1	89.4	53.9
1889	48.1	57.1	84.1	57.2
1899	48.5	60.5	80.1	44.4
1899	48.2	59.2	81.5	53.5
1904	50.4	63.2	79.8	46.3
1909	44.9	69.3	64.7	48.6
1914	41.6	66.7	62.3	53.7
1919	65.1	80.6	80.8	61.7
1921	46.7	59.8	78.2	63.3
1923	51.6	73.3	70.4	58.4
1925	48.1	68.0	70.8	55.5
1927	46.1	66.1	69.9	57.0
1929	47.8	68.9	69.4	61.0
1931	33.1	49.8	66.5	62.1
1933	30.4	46.1	65.9	61.2
1935	38.7	56.6	68.4	64.4
1937	46.3	66.5	69.6	62.1
1939	40.0	59.7	67.0	63.2
1947	53.5	83.0	64.5	61.2
1954	60.7	76.4	79.4	76.2
1958	56.4	67.1	84.0	81.4
1963	50.8	64.9	78.3	78.1
1967	54.5	70.7	77.1	77.5
1972	49.6	64.6	76.7	77.6
1977	51.1	62.3	82.1	79.8
1982	48.7	53.5	91.0	92.3
1987	45.9	50.7	90.6	86.6
1992	36.1	45.6	79.1	84.4
1997	37.0	45.8	80.9	86.2

by two-digit industries until 1959. Official statistics on gross state product began in 1963.

ratio, it implies that the state's relative per capita income was greater than its relative manufacturing output per production worker. This condition typically held in California, leading manufacturing to be underrepresented.

In 1859, manufacturing activity was relatively uncommon in California. There were about 16 manufacturing production workers for every thousand California residents compared with 37 per thousand in the country as a whole. By this measure, California was roughly on a par with the South Atlantic region (which had 17 production workers per thousand residents) but was far less industrialized than New England (which had 114) or the Mid-Atlantic states (with 61).[2]

By 1879, the number of production workers per thousand Californians had doubled to 46, climbing above the national ratio of 1859. But in the intervening 20 years, the national number had increased to 51. At this time, the California/U.S. relative employment ratio stood at around 90 percent (that is, 46/51), whereas the output ratio was a little above 50 percent. Both ratios held steady over the remainder of the 19th century.

These data indicate that manufacturing was less important in 19th-century California than in the nation as a whole. They also provide a hint about the cause—the state's productivity differential in manufacturing was far smaller than its overall income differential. It was not that California manufacturing workers were less productive than their eastern counterparts. In fact, value added per worker in California was always significantly above the national average over this period. It was simply that the relative differentials in other activities in 19th-century California were even more favorable. Indeed, the authors of the state's manufacturing chapter in the *Twelfth Census* observed that before

---

[2]The East North Central states (Ohio, Indiana, Illinois, Michigan, and Wisconsin) had roughly 21 manufacturing production workers per thousand people in 1859 (and 81 by 1899). Data on the number of production workers outside California are from Niemi (1974). These numbers actually understate the difference because the state's population was disproportionately weighted toward prime-age adults in this period. For example, in 1860, those over age 20 made up 71 percent of the California population; the national number was just under 50 percent. This implies that the manufacturing production worker gap in terms of "economically active members of the population" was even greater than indicated (Thompson, 1955, pp. 54–55).

1900: "the geographical position of the state, the high rate of wages, the high price of fuel, and the exceptional attractions offered by mining and agriculture . . . have limited the growth of manufactures in California, and have determined, in a large measure, the particular lines established."[3] The relative weakness of manufacturing activity, which was considered the engine of modern growth over this period, did cause concern among California's political and business leaders about the region's long-run prospects.

As California's population growth accelerated after 1900, manufacturing growth began to increase, but its expansion was not as rapid. Between 1899 and 1914, the number of manufacturing workers per thousand Californians fell from 48 to 42, whereas the national number rose slightly from 63 to 67. Manufacturing was becoming even more underrepresented in the state, which suggests that the industrial sector's expansion, although part of the renewal of growth in the state, was not yet its driving force. The production worker–population ratio moved around erratically between 1914 and 1939, rising to an all-time peak of 65 workers per thousand Californians in 1919 and falling to a 20th-century low of 30 at the bottom of the Great Depression in 1933. In 1939, the ratio was slightly below its 1914 level.

In the two decades after 1939, manufacturing activity began growing much more rapidly than the economy as a whole in the state. The number of manufacturing production workers per thousand Californians rose from 40 in 1939 to 61 in 1954, an increase of over one-half. By 1958, the California/U.S. employment ratio reached 84 percent and the output ratio climbed to 81 percent. Manufacturing began to act as driving force in California's economic expansion over the 1939–1958 period.

After the mid-1950s, the number of production workers per capita fell both nationally and in California, reflecting the declining importance of manufacturing in the overall economy. The decrease was especially rapid nationwide during the late 1970s and early 1980s. Between 1977 and 1987, the number of manufacturing production workers per thousand U.S. population declined from 62 to 51. The fall in the state

[3] *1900 Census*, Part II, *Manufactures*, p. 33.

was less dramatic, dropping only from 51 to 46. As a result, in 1987, the California/U.S. employment ratio reached 91 percent, near its highest level in the Census records. The output ratio also rose into this range in the 1980s. Although manufacturing was still underrepresented in the state, the gap was smaller in the 1980s than ever previously recorded in the *Census of Manufactures*. It is a little ironic that the relative position of manufacturing in California came closest to its national status only when the sector's overall importance was in decline.

4. Structural Transformations in California Manufacturing

Changes in California's manufacturing sector provide valuable signals about the direction that the region's overall economy is moving. The same fundamental forces, such as improvements in transportation and communications, changing factor scarcities, shifts in demand, and growing technological capabilities, as well as the innumerable transitory shocks that shaped the overall California economy, left their mark on the development of its manufacturing sector. Moreover, the major transformations occurring within California's manufacturing sector over the 20th century—specifically, the declining importance of resource-processing activities and the rising importance of information-based activities—are of considerable interest in their own right.

Shifts Between Durable and Nondurable Goods Production

The most common distinction used to understand shifts in industrial composition is that between durable and nondurable goods. (The formal dividing line is whether the goods normally last three years or more.) Durable goods production is traditionally thought to characterize a more mature or advanced economic structure but is also often associated with greater cyclical volatility. The durable goods sector, led by lumber production, initially dominated the California industrial structure, consisting of 60 percent of employment in 1859. By 1879, that figure had fallen to about 37 percent. Over the late 19th and early 20th centuries, the share grew slowly and unevenly until in 1939 it stood at 44 percent. After 1939, the durable share soared. By 1967, over 56 percent of California's production workers were employed in the durables sector. Movements of the sector's share of value added were broadly similar. This profile resembles that for the nation as a whole, although surges in

California's durable goods shares during two periods—1939–1958 and 1977–1987—were more dramatic. Over the past decade, durable goods shares declined slightly. By 1997, the category accounted for 55 percent of California manufacturing value added and 52 percent of production workers.

Consumer and Capital Goods Production

Another key transformation involved the movement from consumer to capital goods production. Early industrialization historically tended to concentrate on producing consumer goods, but over time, output and employment shifted to capital goods industries (Hoffman, 1958, pp. 16, 145). California's experience fits this general pattern well. Indeed, the transition was sharper and more pronounced in the state than in the nation as a whole. Circa 1879, about half of California's industrial activity was devoted to producing consumer goods and less than one-quarter was devoted to capital goods production. Since that date, the share of consumer goods has generally declined. There were two exceptions: during the Great Depression, when falling incomes and sharply reduced investment spending pushed relative demand toward consumer goods, and during the recent period, when the sector's share of production workers in California rose as a result of the expansion of the apparel and related industries.

Capital goods production also follows the expected pattern. In California, the capital goods share was relatively low and constant between 1869 and 1914, rose modestly in the 1920s, and began to climb rapidly after the mid-1930s. By the mid-1950s, California's capital goods share had risen above the corresponding U.S. share. In essence, California's shift toward capital goods production was slower to start, but once initiated, it was sharper and more pronounced.

Resource-Based Manufacturing Activities

One key transition in California's industrial structure over the 20th century was the shift away from resource-based manufacturing activities. Resource-based industries dominated California manufacturing during the late 19th and early 20th centuries, when the fruit and vegetable canning and petroleum refining industries gained prominence. As late as 1935, resource processing still accounted for over half the industrial activity in California. That share declined dramatically after 1939, falling to about one-quarter of industrial activity by 1963. During the mid-1980s, both the employment and output shares of the resource-intensive sectors fell below those prevailing in the country as a whole for the first time in the state's economic history.

In the United States as a whole, the resource-intensive sector was generally characterized by higher levels of labor productivity, especially before World War I and again in the 1930s and 1940s. Labor productivity differentials were less apparent in California, where the employment and output shares were always close. This implies that the productivity gap in California was smaller, suggesting greater integration of the different sectors in the state.

Information Technology

A second key transition in California was the rise of knowledge-based industries, of which the information technology (IT) sector formed an important part (Porat, 1977). Figure 4.1 shows the shares of the IT manufacturing in production workers and value added in California and the United States from 1939 to 1997. In 1939, the information sector share was small nationally and smaller in the state. But with a surge of growth in the late 1950s and early 1960, California passed the country as a whole. And with another surge between 1972 and 1982, the IT sector's share of California value added climbed to 27 percent and production workers to 32 percent. After 1982, the share of production workers started to fall in the state, but the value added share continued to rise, reaching 37 percent by 1997. This pattern contrasts with the stability of the national value added share, which has remained in the mid-20 percent range since 1982.

Factor Intensities

A final useful way to understand California's industrial evolution is to consider shifts of manufacturing activity between capital-intensive, labor-intensive, and high value added industries (Shapira, 1986).

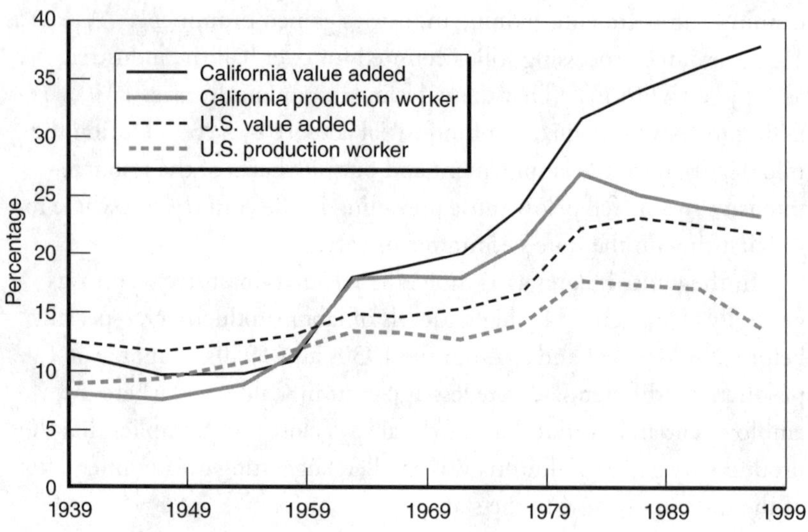

Figure 4.1—Information Sectors in California and the United States, 1939–1997

Capital-Intensive Share

During the early 20th century, capital-intensive industries were considerably more important in California than in the national as whole. Between 1899 and 1939, the capital-intensive sector generally accounted for about 55 percent of employment in California, whereas it made up only about 45 percent nationally. But after 1939, the relative size of the capital-intensive sector declined both in California and in the country as a whole. The relative decline in the state was not the result of an absolute contraction of activity in capital-intensive industries. Rather, it was the product of the more robust expansion of the other parts of California manufacturing and, in particular, of the high-technology sector. Between 1963 and 1977, the capital-intensive shares of California's employment and output remained roughly constant in the mid-30 percent range.

Thereafter, the downward trend resumed, with greater intensity in the state than in the nation. The contraction was in absolute as well as relative terms. The number of production workers in the state's capital-intensive sector fell from 378,000 in 1977 to 362,000 in 1982 (a recession year). Even when the manufacturing sector recovered in the late 1980s, employment in the capital-intensive sector remained below its 1977 levels. By 1997, the capital-intensive sector was about 11 percentage points less important in California than in the nation as a whole, reversing the relative positions of the pre-1939 period.

There is another noteworthy difference between the national and California experiences. Nationally, the capital-intensive sector was always characterized by substantially higher labor productivity than manufacturing as a whole. Not so in California, where labor productivity in the capital-intensive sector was roughly equal to that in other sectors taken as a whole.

Labor-Intensive Share

As one would expect in a historically high-wage region, the labor-intensive industries were consistently less important in California than in the nation as a whole. For example, in 1899, the labor-intensive sector made up 39 percent of employment in California compared with 49 percent nationally. In both areas, the sector's relative position declined up to the mid-1960s, partly because labor productivity in the labor-intensive sector was lower and generally grew more slowly than in manufacturing as a whole. After 1967, the employment share of the labor-intensive sector in California increased by over 10 percentage points (compared with a 6 point increase nationally). Much of the rise occurred between 1967 and 1977, when the state's high-technology sector grew relatively slowly. Employment gains in the state's labor-intensive sector have proved long-lasting. By 1997, the employment share in the state was higher than at any point since World War I.

High-Technology Share

During the 1990s, the high-technology sector accounted for about one-half the manufacturing value added in California and over one-quarter of the employment of production workers. By way of contrast, the sector made up about one-third of output and less than one-fifth of employment nationally. Over the 1909–1939 period, California began to specialize in knowledge-intensive activities, as high-technology industries grew more rapidly in the state than in the nation. By 1939, the high-technology sector accounted for 21 percent of output and 14 percent of production workers in California. These shares were 4–5 percentage points higher than in the country as a whole.

The high-technology sector's "takeoff" occurred between 1939 and 1967. Expansion was disproportionately vigorous in the state, driven initially by the growth of its aircraft industry and later by electronics. The early 1970s witnessed a brief interruption in this story of endless expansion. Between 1967 and 1972, the share of the high-technology sector in California manufacturing output fell by about 6 percentage points and its share of production workers declined by almost 12 percentage points. The downturn in the state was of much greater relative magnitude than was the national contraction.

During the late 1970s and early 1980s, the sector bounced back. This surge, which was linked to the growth of the semiconductor and computer equipment industries, impressed many contemporaries as the beginning of a new era in the state's economic development. But the long-run data make it clear that the 1977–1982 episode is better viewed as the upswing in a long cycle rather than as a breakpoint in a growth trend. The largest change in the high-technology share in California occurred between 1947 and 1954, when this sector's share of the state's manufacturing value added climbed from 24 percent to 37 percent and its share of production workers rose from 22 percent to 34 percent. These figures indicate that the high-technology sector in California broke away from the rest of the country during the late 1940s and 1950s, not during World War II or the Carter-Reagan years.

Changes in the state over the 1967–1997 period can be summarized as follows: The labor-intensive sector, where labor productivity was low, experienced a growing employment share and a roughly constant output share; the high-technology sector, where productivity was high, had both growing employment and output shares; and the capital-intensive sector, where productivity was closer to the average, experienced declining output and employment shares. This picture mirrors the popular perception of a "disappearing middle," of an industrial structure increasingly polarized into workplaces with high-productivity, high-paying jobs and those with low-productivity, low-paying jobs. Although

California has followed the national trends in this regard, it has done so with greater intensity. These developments are unprecedented in California's industrial evolution, and the policy challenges they present are new.

5. California's Changing Comparative Advantages

A quick and insightful way to contrast the manufacturing sectors in California and the United States is to compare their top-ten lists. Two lists are instructive. The first includes the industries that accounted for the most industrial activity in the state and in the nation. The second lists those industries, however large or small, in which California accounted for the largest share of U.S. activity. This helps illustrate the state's changing comparative advantages.

Measured by Size

Although California's industrial structure broadly resembled that of the United States throughout the 20th century, there remained significant differences in specialization at the four-digit industry level. Table 5.1 lists the top-ten four-digit industries, as ranked by manufacturing value added, in the state and the country as a whole in 1869, 1899, 1929, 1963, 1977, and 1997. Focusing first on the national picture, we observe a transition over the late 19th and early 20th centuries from a list dominated by light consumer goods such as textiles and shoes to one including many heavy capital goods such as machinery and iron/steel. In the early 20th century, motor vehicles rose to the top of the national list and by the 1950s, aircraft and radio and TV communications equipment broke into the national top ten.[1]

Although California's top-ten list consistently included many of the same industries—lumber, newspapers, and bakery products—as its national counterpart, the differences are noteworthy. Textiles and iron/steel products, which were staples on the national list in the late

[1]In 1987, the Census reclassified most radio and TV communications equipment as search and navigation equipment.

Table 5.1

The Top-Ten Industries in California and the United States Measured by Value Added and Production Workers

Rank	California	United States
	Value Added Measure	
	1869	
1	Lumber, sawed	Lumber, sawed
2	Flour and grist mill products	Boots and shoes
3	Boots and shoes	Flour and grist mill products
4	Tobacco and cigars	Cotton goods
5	Newspapers	Clothing, men's
6	Steam engines	Woolen goods
7	Machinery	Iron forging
8	Malt liquor	Carriages and wagons
9	Carriages and wagons	Iron castings
10	Quicksilver, smelted	Furniture
	1899	
1	Lumber and timber products	Foundry and machine shop
2	Foundry and machine shop	Lumber and timber products
3	Newspapers	Iron and steel work
4	Canned fruits and vegetables	Liquor, malt
5	Malt liquor	Newspapers
6	Smelting and refining, copper	Cotton goods
7	Shipbuilding, iron	Clothing, men's
8	Planning mill products	Tobacco, cigars, and cigarettes
9	Copper, tin, sheet-iron work	Slaughtering and meat packing
10	Bread and bakery products	Boots and shoes
	1929	
1	Petroleum refining	Foundry and machine shop
2	Canned fruits and vegetables	Iron and steel, rolling mills
3	Newspapers	Newspapers
4	Foundry and machine shop	Electrical machinery apparatus
5	Motor vehicles	Motor vehicles
6	Lumber and timber products	Lumber and timber products
7	Bread and bakery products	Bread and bakery products
8	Printing and publishing, book	Clothing, women's
9	Furniture	Printing and publishing, book
10	Rubber tires	Tobacco, cigars, and cigarettes
	1947	
1	Aircraft	Motor vehicles and parts
2	Petroleum refining	Rolling mills
3	Canned fruits and vegetables	Cotton broad woven goods
4	Sawmills	Petroleum refining

Table 5.1 (continued)

Rank	California	United States
5	Newspapers	Sawmills
6	Motor vehicles and parts	Newspapers
7	Bread and bakery products	Bread and bakery products
8	Wineries	Paper and paperboard mills
9	Shipbuilding	General commercial printing
10	General commercial printing	Meatpacking plants
	1963	
1	Ordnance and accessories	Motor vehicles and parts
2	Radio and TV communications eq.	Blast furnaces and steel mills
3	Aircraft	Radio and TV communications eq.
4	Aircraft engines and parts	Aircraft
5	Motor vehicles and parts	Newspapers
6	Aircraft equipment	Petroleum refining
7	Petroleum refining	Ordnance and accessories
8	Newspapers	Organic chemicals
9	Canned fruits and vegetables	Pharmaceutical preparations
10	Bread and related products	Bread and related products
	1977	
1	Radio and TV communications eq.	Motor vehicles
2	Guided missiles	Motor vehicle parts
3	Aircraft	Blast furnaces
4	Petroleum refining	Petroleum refining
5	Electronic computing equipment	Plastic products
6	Semiconductors	Industrial organic chemicals
7	Newspapers	Radio and TV communications eq.
8	Aircraft parts and equipment	Newspapers
9	Canned fruits and vegetables	Pharmaceuticals
10	Sawmills	Aircraft
	1997	
1	Electronic computers	Semiconductors
2	Semiconductors	Pharmaceuticals
3	Telephone apparatus	Light truck and utility vehicles
4	Radio and TV communications eq.	Plastic products
5	Petroleum refineries	Newspaper publishers
6	Guided missiles and space vehicles	Petroleum refineries
7	Search and navigation equipment	Commercial printing
8	Newspaper publishers	Automobiles
9	Plastic products	Electronic computers
10	Instruments for testing electric devices	Iron and steel mills

Table 5.1 (continued)

Rank	California	United States
	Production Worker Measure	
	1977	
1	Communications equipment	Motor vehicle parts
2	Aircraft	Plastic products
3	Electronic computers	Blast furnaces
4	Canned fruits and vegetables	Motor vehicles
5	Guided missiles	Communications equipment
6	Industrial machinery	Commercial printing
7	Aircraft parts and equipment	Sawmills
8	Dresses	Industrial machinery
9	Electronic components	Newspapers
10	Sawmills	Dresses
	1997	
1	Women's, girls', and infants' cut and sewn apparel contractors	Broadwoven fabric finishing mills
2	Plastic products	Curtain and drapery mills
3	Machine shops	Plastic products
4	Commercial printing	Commercial printing
5	Semiconductors	Machine shops
6	Motor vehicle parts	Poultry processing
7	Aircraft	Motor vehicle parts
8	Printed circuit assembly	Newspaper publishers
9	Bare printed circuit board	Women's, girls', and infants' cut and sewn apparel contractors
10	Sheet metal work	Animal slaughtering

19th and early 20th centuries, were essentially no-shows in California. In this sense, California's experience did not fit the traditional industrialization model pioneered by Britain and followed by the early developing regions of Europe, the eastern United States, and Japan. The absence of cheap labor and suitable coking coal in California caused the leading industries of the "first industrial revolution" to be virtually absent from the region. In addition, many key sectors of the so-called "second industrial revolution"—chemicals, electrical machinery, and automobiles—played much smaller roles in California than in other industrializing regions.

Instead of relying on the traditional "leading sectors," California leapfrogged the older industrial regions to adopt a new and relatively distinct set of activities. In the early 20th century, California's top-ten list was headed by canning, nonferrous metals, and petroleum industries, all of which processed the state's natural resources. Canning remained among the state's top ten largest industries from 1899 into the 1990s. Petroleum refining, which first entered the list in 1914, topped the chart by 1929. It stayed on the list during the remainder of the 20th century, even after California became a large-scale oil importer. Beginning in the 1910s, motor vehicles appeared on both the state and national lists, but the industry's relative importance was generally lower in California than in the nation as a whole. By the 1970s, automaking dropped off the California list while remaining the most important industry nationally. Blast furnace and steel mills represented another sector that was typically close to the top of the U.S. charts over the 20th century but did not rank nearly so high in California. In the place of autos and steel industries, California increasingly focused on aircraft and electronics in the post-World War II period.

Only a few four-digit industries have contributed so significantly to California's growth, through their own expansion and their spillover effects, that they merit the title "leading industry." One prime candidate in the early period is canning, which accounted for about 14 percent of total employment growth in California between 1879 and 1929. At its peak in 1935, it accounted for about 15 percent of California's manufacturing production workers. A candidate for the later period is the aircraft industry, which made up about 11 percent of employment growth between 1939 and 1963 and employed 22 percent of the state's production workers at its peak in 1954. Perhaps the largest one-time contribution to growth was the expansion of the shipbuilding industry between 1914 and 1919, when it accounted for 46 percent of total employment growth. In the 1940s, as well, shipbuilding expanded during the war and then virtually disappeared. Its direct effects were generally short-lived, although some of the effects of the wartime booms, especially their stimulus to the development of the western steel industry, were more sustained.

Measured by Specialization

A second way to gain perspective on the evolution of California manufacturing is to consider the industries in which the state has been most highly specialized, those industries in which the state's share of national value added was greatest. Table 5.2 provides the top-ten lists of California's most specialized four-digit industries in 1869, 1899, 1929, 1947, 1963, 1977, and 1997.[2] As a broad generalization, the 19th-century lists are dominated by industries producing inputs for and processing the output of California's distinctive extractive activities—gold mining and fruit farming. For example, heading the list in 1869 was quicksilver processing, which produced mercury for gold miners.[3]

The star performer over the entire period was winemaking. In 1869, California accounted for almost 40 percent of national value added. By 1899, this fraction had risen to one-half and by 1929 (during Prohibition) to almost three-quarters. (The production of sacramental wine was still permitted, which allowed a handful of California wineries to operate legally.) But even after the repeal of Prohibition, California continued to account for the majority of national value added. Also consistently high on the lists are industries involved in canning and preserving fruits and vegetables. Petroleum refining makes a notable appearance in 1929, when California accounted for over one-fifth of national output.

During the mid-20th century, a set of high-technology goods, led by aircraft, entered the list. In 1947, the state's aircraft and aircraft equipment industries produced about one-half of national output. By 1963, ordnance (principally guided missiles) and electronics began to appear prominently. In 1997, the list comprised a collection of specialized

[2]In constructing the list, I generally excluded industries subject to Census nondisclosure rules. This omits a handful of highly distinctive California industries, but using my estimated activity levels in the rankings of individual industries would add more noise than music.

[3]As the table shows, the state's share of national value added was greater than one. This is not an error. The value added of the industry outside California was negative, which happens when the cost of materials exceeds the value of the product. Negative value added leaves nothing to cover labor, capital, and managerial costs and is obviously not sustainable over the long run.

Table 5.2

The Top-Ten Most Specialized Industries in California Measured by California's Share of National Value Added

Rank	Industry	Share
	1869	
1	Quicksilver, smelted	4.16
2	Liquor, vinous	0.39
3	Salt, ground	0.21
4	Explosives	0.17
5	Boxes, cigar	0.10
6	Canned fruits and vegetables	0.08
7	Glue	0.08
8	Hand stamps	0.08
9	Safes and vaults	0.07
10	Bellows	0.07
	1899	
1	Sugar, beet	0.50
2	Liquor, vinous	0.49
3	Wool scouring	0.34
4	Explosives	0.24
5	Canned fruits and vegetables	0.15
6	Shipbuilding	0.09
7	Iron and steel doors	0.07
8	Bags, other than paper	0.07
9	Babbitt metal and solder	0.07
10	Smelting and refining, copper	0.07
	1929	
1	Liquor, vinous	0.72
2	Canning and preserving fish	0.37
3	Oils, not elsewhere classified	0.34
4	Canned fruits and vegetables	0.28
5	Peanuts and other processed nuts	0.26
6	Petroleum refining	0.20
7	Ivory and bone work	0.20
8	Rice cleaning	0.16
9	Roofing materials	0.12
10	Steel barrels	0.12
	1947	
1	Wineries	0.73
2	Aircraft equipment	0.60
3	Dehydrated fruits and vegetables	0.55
4	Canned and cured seafood	0.50
5	Aircraft	0.48

Table 5.2 (continued)

Rank	Industry	Share
6	Canned fruits and vegetables	0.26
7	Oil-field machinery and tools	0.23
8	Women's and misses' outerwear	0.22
9	Rice cleaning	0.21
10	Auto trailers	0.21
	1963	
1	Wineries	0.67
2	Ordnance and accessories	0.60
3	Dehydrated fruits and vegetables	0.58
4	Electron tubes, transmitting	0.45
5	Canned and cured seafood	0.37
6	Rice cleaning	0.36
7	Canned fruits and vegetables	0.31
8	Electrical measuring instruments	0.26
9	Aircraft equipment	0.25
10	Aircraft engines and parts	0.25
	1977	
1	Wineries	0.79
2	Guided missiles and space vehicles	0.72
3	Fur goods	0.66
4	Dehydrated fruits and vegetables	0.52
5	Canned and cured seafood	0.48
6	Space propulsion units and parts	0.48
7	Fine earthenware	0.33
8	Canned fruits and vegetables	0.31
9	Motor homes	0.31
10	Aircraft	0.31
	1997	
1	Wineries	0.88
2	Guided missiles and space vehicles	0.57
3	Magnetic and optical recording media	0.50
4	Dried and dehydrated food	0.50
5	Women's and girls' cut and sewn blouses and shirts	0.48
6	Space propulsion units and parts	0.45
7	Electronic computers	0.45
8	Instruments for testing electric devices	0.42
9	Semiconductor machinery	0.41
10	Telephone apparatus	0.40

resource-processing industries (wineries and dried/dehydrated food), high-technology industries (guided missiles, electronics, telecommunications equipment, and instruments), and certain branches of the apparel industry. In each case, California accounted for over 40 percent of national value added. (Just below the cut, in 11th place, was tortilla manufacturing, an industry separately enumerated in the Census for the first time.)

Changes in California's Comparative Advantages

A final, more systematic way to explore the changing nature of the state's comparative advantages is to examine shifts in specialization across broad two-digit sectors. Here the best measure is the ratio of the industry's share of state manufacturing value added relative to its share of national value added. (The ratio is also known as the industry's location quotient or coefficient.)[4] Table 5.3 shows the changing ratio of relative shares for 1859, 1899, 1939, 1967, 1987, and 1997. If an industry's ratio is greater than one—it constituted a larger fraction of industrial output in the state than in the nation—the activity is "overrepresented" in California. The state is said to have a comparative advantage in such industries. If the ratio is less than one, it is "underrepresented" and the state has a comparative disadvantage.

During the Gold Rush era, California's pattern of comparative advantage was quite distinct. Lumber production led the way in 1859 with a share of California's manufacturing output that was almost three times its share nationally. Food processing followed closely, with a California share more than twice its national share. Among the other industries overrepresented in the state were printing, transportation equipment, and nonelectrical machinery. All other industries were underrepresented in 1859.

This general pattern of specialization continued over the late 19th century. But there were changes in the rankings and some regression

---

[4]The location quotient or coefficient is the percentage of state activity in a given industry divided by the percentage of national activity in a given industry.

Table 5.3

Ratio of California to U.S. Manufacturing Shares Measured by Value Added and Production Workers

Rank	Industry	1859	1899	1939	1967	1987	1997
	Value Added Measure						
20	Food	2.55	1.71	1.82	1.26	1.12	1.07
21	Tobacco	0.24	0.37	0.14	--	--	0.05
22	Textiles	0.10	0.14	0.19	0.15	--	0.32
23	Apparel	0.18	0.78	0.71	0.62	1.08	1.63
24	Lumber	2.98	1.61	1.70	1.20	0.87	0.66
25	Furniture	0.36	0.61	1.35	1.03	1.10	0.98
26	Paper	0.21	0.35	0.56	0.57	0.49	0.46
27	Printing	1.27	1.44	1.23	0.91	0.93	0.97
28	Chemicals	0.46	1.24	0.89	0.53	0.40	0.48
29	Petroleum/coal	—	0.34	2.77	1.14	1.38	1.44
30	Rubber/plastics	—	0.19	1.22	0.86	0.84	0.79
31	Leather	0.42	1.06	0.22	—	—	0.66
32	Stone/clay/glass	0.62	0.85	1.09	0.89	0.93	0.78
33	Primary metals	0.40	0.71	0.49	0.52	0.37	0.35
34	Fabricated metals	0.68	0.69	1.14	0.96	0.87	0.88
35	Nonelectrical machinery	1.15	0.97	0.70	0.72	1.17	1.32
36	Electrical machinery	—	0.23	0.43	1.38	1.51	1.81
37	Transportation equipment	1.22	1.62	1.26	1.45	1.39	0.73
38	Instruments	—	0.47	0.34	0.83	1.66	1.81
39	Miscellaneous manufacturing	0.38	0.58	0.50	0.91	0.89	1.15
19	Ordnance	—	—	—	5.34	—	—
	Production Worker Measure						
20	Food	3.00	2.89	2.49	1.34	1.11	1.11
21	Tobacco	0.23	0.59	0.31	—	—	0.11
22	Textiles	0.09	0.14	0.12	0.11	—	0.36
23	Apparel	0.11	1.03	0.84	0.65	1.12	1.90
24	Lumber	3.62	1.58	1.57	1.09	0.91	0.70
25	Furniture	0.31	0.54	1.49	1.03	1.24	1.14
26	Paper	0.24	0.42	0.60	0.67	0.59	0.63
27	Printing	1.68	1.58	1.41	1.06	1.01	1.02
28	Chemicals	0.59	1.19	1.01	0.61	0.57	0.72
29	Petroleum/coal	—	0.18	2.53	1.53	1.24	1.22
30	Rubber/plastics	—	0.20	1.10	0.94	0.92	0.88
31	Leather	0.41	0.96	0.21	—	—	0.95
32	Stone/clay/glass	1.55	0.64	1.13	0.95	0.92	0.87
33	Primary metals	0.95	0.50	0.56	0.61	0.49	0.43
34	Fabricated metals	0.80	1.05	1.21	1.06	0.98	0.98
35	Nonelectrical machinery	1.11	0.74	0.72	0.84	0.95	0.91

Table 5.3 (continued)

Rank	Industry	1859	1899	1939	1967	1987	1997
36	Electrical machinery	—	0.34	0.45	1.31	1.41	1.52
37	Transportation equipment	1.67	1.61	1.45	1.65	1.25	0.79
38	Instruments	—	0.31	0.38	1.09	1.78	1.61
39	Miscellaneous manufacturing	0.35	0.58	0.48	0.93	0.98	1.23
19	Ordnance	—	—	—	4.17	—	—
	Ratio of population to production workers	2.29	1.25	1.49	1.30	1.10	1.20

toward the mean.[5] In 1899, food processing had the highest degree of specialization, followed by transportation equipment, lumber products, printing, chemicals, and leather goods. All other industries were underrepresented. Again, one striking feature was the nearly complete absence of a textile industry in the state. This industry made up 9.5 percent of national industrial value added in 1899 but only 1.3 percent in California.[6] Primary metal production, which included the iron and steel industry, was also relatively uncommon in the state.

By 1939, the state had added several new specialties. Petroleum products, which had been underrepresented in 1899, became highly overrepresented on the eve of World War II. Rubber products also shifted from being an underrepresented to an overrepresented industry. Following traditional patterns, food, printing, transportation, and fabricated metals were overrepresented; tobacco, textiles, leather, instruments, and primary metals were highly underrepresented.

The headlines in 1967 were

- The increasing specialization in electrical equipment; by 1967, its share of state manufacturing activity was more than one-half higher than its share of national activity;

[5]That is, industries with high relative shares of manufacturing activity in 1859 saw declines and those with low relative shares saw increases by 1899.

[6]But note that if the state's employment were compared with national population rather than manufacturing employment, these industries would no longer appear "overrepresented."

- The ascendance of ordnance and transportation equipment (led by aircraft production) to the status of the sectors with the greatest relative strength; and
- The weakening of the state's traditional specialization in resource-based sectors such as petroleum, lumber, and food processing.

California's development over the 1970s and 1980s witnessed the growing specialization in apparel. The state also solidified its specialization in the block of sectors including machinery (both electrical and nonelectrical), transportation equipment, and instruments. The shares of these sectors in California's output were about one-half again as high as they were nationally. Tobacco, textiles, primary metals, leather, chemicals, and paper products were underrepresented to a similar or greater extent. But what is notable is how close most of the ratios were to unity—by 1987 California's industrial structure had become quite similar to that of the United States as a whole.

The most important development between 1987 and 1997 was the shift of transportation from the status of a perennially overrepresented industry to an underrepresented industry. Obviously, this shift was due to the contraction of the state's aircraft sector following the post–Cold War military cutbacks and represented a notable reversal of one of California's historical patterns of specialization. The second notable development was the further growth of California's comparative advantage in apparel. By 1997, the industry had become one of the most overrepresented sectors in the state. In one sense, the emergence of an apparel sector was an outgrowth of deeply rooted developments in the state's industrial history. In another important sense, it represented a marked departure from California's traditional weakness in "low-wage" manufacturing. Many traditional patterns continued over the past decade. Electrical equipment, instruments, petroleum, and food processing remained areas of relative strength. Tobacco, textiles, primary metals, lumber, paper, and chemicals remained areas of relative weakness.

All of these changes in California's comparative advantages took place within an industrial economy that was generally expanding rapidly over the 20th century. Absolute decline was rare. Historically, California's manufacturing sector has been characterized by balanced growth, with expansion virtually across the board.

6. California Manufacturing as a Pace-Setter

In addition to outpacing the nation as a whole in manufacturing over the 20th century, California's industrial development anticipated and stimulated many national trends. This chapter identifies and discusses seven areas in which California manufacturing led the way.

The Shift to Higher Value Added Activities

A higher value added ratio is often taken as a sign of a more advanced industrial sector, with greater concentration on producing sophisticated goods and less on resource processing. (Matthews, 1996). Figure 6.1 graphs the share of value added in the aggregate value of manufacturing products in California and the United States from 1899

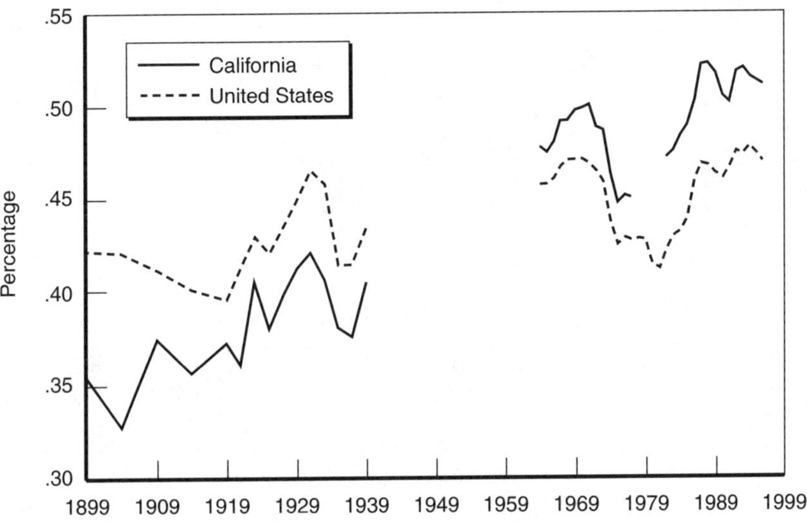

Figure 6.1—Value Added as a Share of Production

to 1997. As the figure shows, the national ratio climbed from 42 percent to 47 percent during this period. Over the same period, the California ratio rose from 35 percent to 52 percent, or more than three times as rapidly. This rise reflects the shift of the manufacturing distribution to higher value added industries and a corresponding decrease in the dependence on natural resources. This growth in high value added activities is one important way that California outpaced the national changes.

An Incubator for New Industries

Another way that California set the pace for 20th-century manufacturing was by becoming an incubator for new, modern industries. Although California has long been a highly innovative place, it has not always been a location where new industries flourished. Figure 6.2 shows the average "age" of industries in California and the United

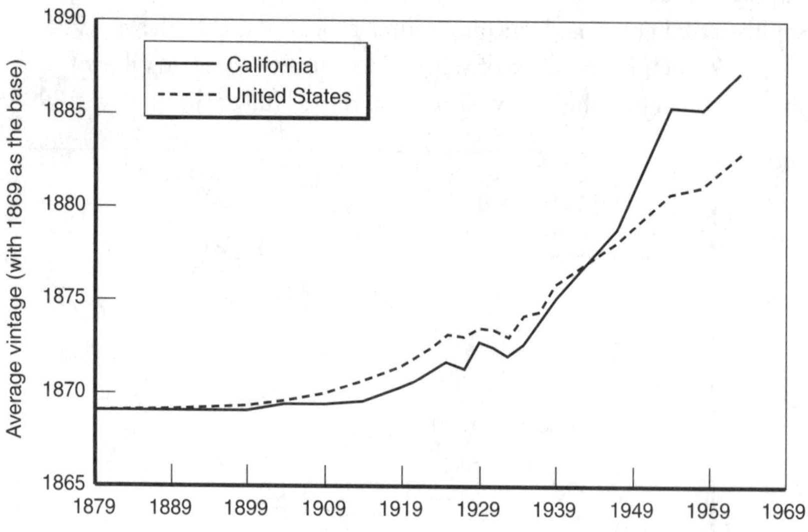

Figure 6.2—Average Vintage of Industrial Composition in California and the United States, 1879–1963

States as a whole between 1879 and 1963.[1] As the figure shows, new industries were slightly less prevalent in California over the late 19th century, and the age gap widened over the 1910s and 1920s. By 1927, the gap reached almost two years. But over the late 1930s, the gap closed and during World War II, new industries became relatively more prevalent in the state than in the nation as a whole. The pattern had become far more pronounced by the mid-1950s. Although it is tempting to highlight World War II as the key crossing point, California entered the passing lane in the 1930s.

A similar picture emerges if one focuses on the relative importance of new industries. Consider three sets of new manufacturing activities—those that were infant industries in the pre–World War I period, (1899–1914), in the interwar years (1919–1939), and the immediate post–World War II period (1947–1958).[2] Figure 6.3 compares the relative prevalence in the new industries in California. Taken collectively, the new industries of pre–World War I period were always less important in California than in the nation as a whole. The interwar group, which

---

[1]The series are calculated by assigning a "birthday" to each four-digit industry (typically the year when the industry first appeared in the *Census of Manufactures*) and then weighing up the industries by their value added. Industries in existence at the beginning of the sample period were assigned an 1869 birthday. A larger number here implies a "younger" industrial mix.

[2]The pre–World War I category includes SIC 3011 rubber tires; 3411 tin ware; 3651 phonographs; 3699 electrical machinery, apparatus, and supplies; and 3710 motor vehicles. The new industries of the interwar period include 2221 and 2241 rayon textiles; 2823 rayon and allied products; 3361 aluminum manufactures; 3522 tractors; 3585 refrigerators; 3651 radio (and later TV) receiving sets; 3581 laundry equipment; and 3720 aircraft and parts. The new industries of the post–World War II period include 2037 quick-frozen foods; 2432 plywood mills; 2821 plastic materials; 2822 synthetic rubber; 3079 plastic products; 3571 computing and related machines; 3662 radio and TV communication equipment; 3670 electronic components and semiconductors; and 3999 ordnance and accessories (chiefly guided missiles). Note that the new industries are not always high-technology, even by the standards of the periods of their "infancy." In some instances, the Census began enumerating an industry before it became sufficiently important to enter my list. As an example, when plywood mills were first enumerated in 1939, they were of minor significance compared with their postwar status. Hence, I treat this sector as a new post–World War II industry.

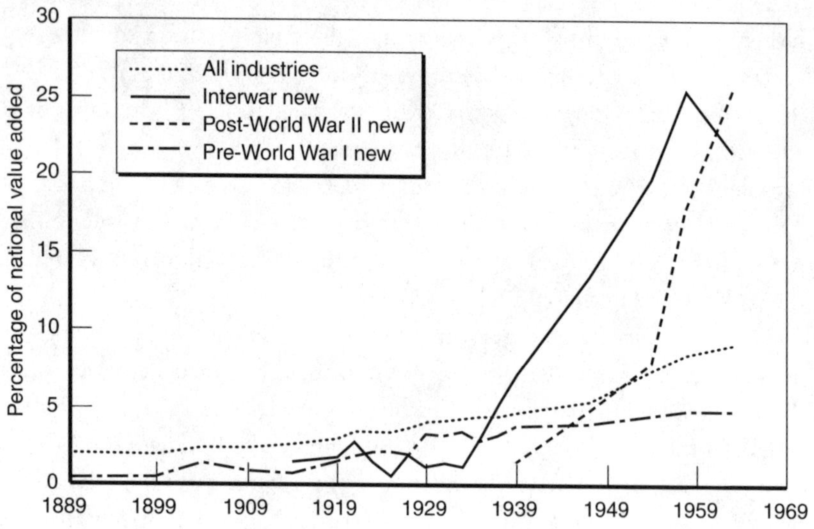

Figure 6.3—California's Share of Value Added of "New" Industries, 1889–1963

includes aircraft, was also initially underrepresented in the state. But after 1935, especially after 1947, this group became significantly overrepresented in California. And the postwar group was overrepresented almost from birth.

Adoption of Modern Power Sources

California's manufacturing sector was also quick to adopt modern sources of industrial power. In the first quarter of the 20th century, the state's manufacturers were among the global leaders pioneering the use of petroleum and natural gas for industrial applications. They also led the way in shifting to electric power generated outside the plant, with important consequences for productivity and factory organization.

Over the late 19th century, manufacturing in California was far less power-intensive than in the nation as a whole (Table 6.1). In 1889, for example, there were about 128 units of primary horsepower capacity for each production worker in California compared with 160 units for the representative wage-earner in the nation. This difference was largely due to the absence of high-quality coal in the state and the need to import

Table 6.1

Aggregate Horsepower per 100 Production Workers, 1889–1962

Year	United States	California	Ratio
1889	160	128	0.80
1899	207	171	0.83
1904	260	223	0.86
1909	291	304	1.05
1914	330	367	1.11
1919	336	328	0.98
1925	437	455	1.04
1929	491	541	1.10
1939	628	680	1.08
1954	964	821	0.85
1962	1,249	1,277	1.02

this fuel at prices two to four times the levels prevailing in the major industrial centers. As noted above, it was commonly observed in the late 19th century that California could not develop industrially without cheaper fuel (Williams, 1997, pp. 52, 143–147).

California's energy situation changed dramatically after 1900 with the discovery and exploitation of huge oil fields in the southern San Joaquin Valley and Los Angeles region and the extensive development of the state's hydroelectric resources.[3] Over the early 20th century, the amount of horsepower per worker grew rapidly in almost every major industry in California. In 1909, the horsepower-to-worker ratio in California stood at 304 units and the power intensity of the state's manufacturing sector surpassed that of the nation as a whole.[4] By 1929, the power intensity ratio in California stood at 541, 10 percent above the national average and more than three times higher than that prevailing in the state in 1899.

Given the scarcity of local coal, California's shift to petroleum and natural gas represented a crucial adaptation to the region's distinctive

[3]California State Mining Bureau (1914).

[4]For the links between power intensity and scale economies, see Atack (1987) and Cain and Paterson (1986).

resource endowment. The state's industrial firms used oil as fuel well in advance of their eastern counterparts. Indeed, when the *Census of Manufactures* first provided a detailed breakdown of fuel use by energy source in 1909, fuel oil accounted for 92 percent of the British thermal units consumed directly as fuel by California manufacturers. Coal and coke made up only 7 percent. This compares with national shares of 4 percent for fuel oil and 92 percent for coal and coke. California manufacturers were also earlier consumers of natural gas. For example, natural gas constituted almost 33 percent of the fuel directly consumed by the state's manufacturers in 1929 and 58 percent in 1939. Nationally, the fractions were 18 percent in 1929 and 29 percent in 1939, about one-half as high.

More important, perhaps, was the shift of the state's manufacturers to electric power, especially power generated outside the plant. Figure 6.4 shows the percentage of aggregate power derived from electric motors driven by purchased energy in California and United States manufacturing from 1899 to 1962. Both series fit the classical S-shaped diffusion curve. As the figure shows, use of this new power source began

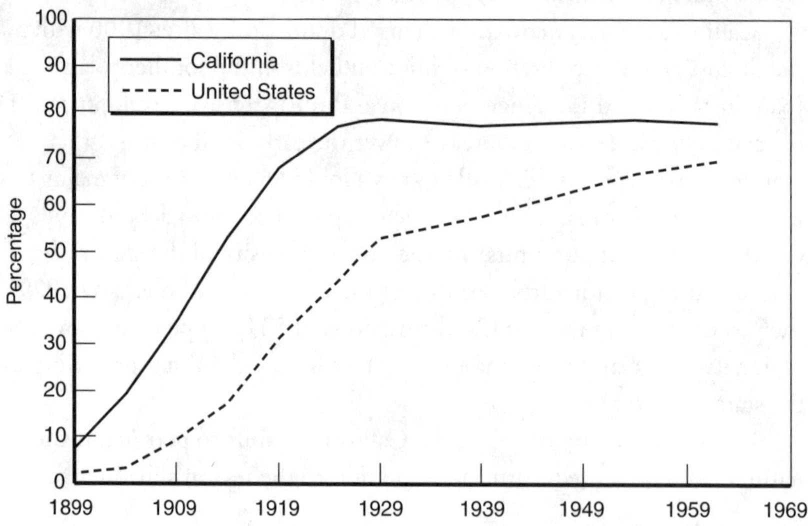

Figure 6.4—Percentage of Aggregate Power from Electric Motors Driven by Purchased Energy, 1899–1962

earlier and spread faster in California than in the nation as a whole. As early as 1899, this source accounted for about 7.6 percent of power used in the state's manufacturing sector compared with about 1.8 percent nationally. The diffusion curve reached over 50 percent in the state by 1914, 15 years ahead of the nation as a whole. By 1929, the California share had already attained its ceiling level of around 80 percent, which was higher than the national curve reached when the available data end in 1962.

As Devine (1983) notes, the shift to electricity in manufacturing, and especially the adoption of fractional reserve engines driven by purchased power, had important consequences for the organization of factories, the skill composition of the labor force, and the scale of production. In factories using steam engines, difficulties in transmitting power led to use of multistoried structures and the concentration of the operations requiring substantial amounts of power near the central drive shaft. Assembly and similar activities were performed elsewhere in the plant and many operations requiring only small amounts of power were done by hand. This often meant that goods-in-process flowing through the factory followed highly circuitous paths. Handling materials in this manner typically used large quantities of unskilled labor (the man with the cart or wheelbarrow).

The introduction and spread of electric motors allowed an entire reorganization and rationalization of the production process. Power, even in small amounts, could be applied wherever and whenever needed. This not only saved labor directly, it allowed the adoption of straightline assembly practices and material-handling techniques, such as conveyor belts and overhead rails. These changes were associated with dramatic increases in labor productivity and a shift in labor demand from unskilled workers to semiskilled or operative workers during the 1920s. The movement to use purchased power also reduced pressures to increase plant size, which were previously driven in part by scale economies in the technologies for generating steam power. Even the physical appearance of the factory changed as extensive, relatively flat structures replaced the multistoried factories of old.

California's industrial takeoff during the early 20th century coincided with a period of relatively low fuel and power costs. Data on rates per kilowatt-hour for industrial power from the *Census of Manufactures* and the *Census of Electric Industries* show that electricity costs in California were consistently 2 to 6 percent *below* the national average in the years between World Wars I and II when the key changes were taking place.[5] The lower energy costs of this period contrast sharply with the pre-1900 period, when energy costs were significantly higher in the state. This period of relative abundance proved transitory, coming to an end during the second half of the 20th century. With the growth of local demand, the relative decline of petroleum production in the state, and the use of low-cost hydroelectric sites, California became an importer of energy. Between the mid-1950s and the mid-1970s, industrial electricity rates in the state ranged between 6 and 22 percent above the national average. (Despite the differences in relative trends, electricity rates in both California and the United States were falling significantly in real terms in the period before 1973. As an example, the California industrial rate in 1972 was only 30.5 percent of that charged in 1929.) After the mid-1970s, electric rates began to rise more rapidly in the state than in the nation. By 1992, California manufacturers were paying 68 percent more for electric power than the average for the United States as a whole. Although California has returned to the status of a region with relatively expensive fuel and power, it is important for the state's overall development that its manufacturing sector was near the forefront of a key set of developments in the use of energy over the early 20th century.

The Rising Ratio of Nonproduction to Production Workers

California manufacturers also outpaced the nation in their extensive use of "white collar" labor. As Figure 6.5 shows, the share of

---

[5]Supplementing the *Census of Manufactures* data is information from the *Census of Electricity* reports on central electric plants for 1917, 1922, 1927, 1932, and 1937 and from the *Annual Survey of Manufactures* for inter-Census years during the post–World War II period.

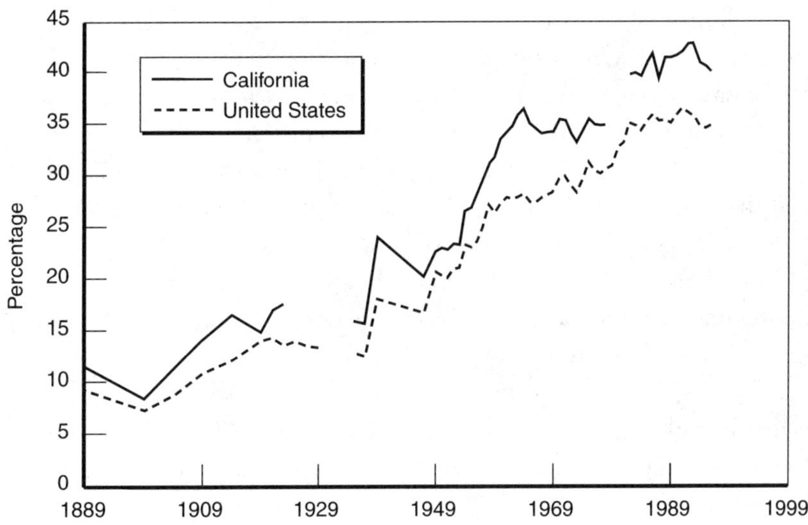

Figure 6.5—Nonproduction Workers as a Share of Total Employment, 1889–1996

nonproduction workers in total employment has been significantly higher in California than in the nation as a whole.[6] This share has generally risen both in California and nationally, but California has typically achieved any given share about a decade before the country as a whole. In his effort to explain the rising ratio of nonproduction to production workers in U.S. manufacturing, Delehanty (1968) found that in a cross-section of industries, higher ratios were correlated with higher capital intensities, research intensities, and most powerfully with higher wages and value added per production worker. In addition, the ratio of nonproduction to production workers tended to fall as the size of plant operations increased. Historically, California fit this pattern closely.

---

[6]These patterns are discussed at length in California Division of Labor Statistics and Research (1967), which found that a "considerable share of California's expansion in manufacturing during the past 2 decades occurred in such 'new' industries as electronics and ordnance (missiles)—industries that have been among the fastest growing and among those with the highest share of white-collar workers" (p. 3). The unusually high ratios in the state and nation in 1939 may be due to changes in the Census enumeration of wage-earners working off the shop floor.

California's industrial structure was generally biased toward industries that had high ratios of nonproduction to production workers in the nation as a whole. The data in Figure 6.6 compare the state and national ratios of nonproduction to production worker with a hypothetical ratio calculated using the national ratio for each industry and the California industrial distribution of production workers over the 1889–1963 period. The hypothetical values using the California industrial distribution exceed the actual national values over most of the period under consideration. This analysis reveals that about half the gap in the ratios of nonproduction to production workers arose because California's industrial distribution was more heavily weighted toward activities with higher ratios of white collar to blue collar workers. It follows that the other half was due to differences within individual industries.

The higher ratio of nonproduction to production workers supports a generalization about California's traditionally higher levels of human capital. Given that average salaries per nonproduction employee were almost always higher than wages per production worker and that the ratio of nonproduction to production workers was higher in California

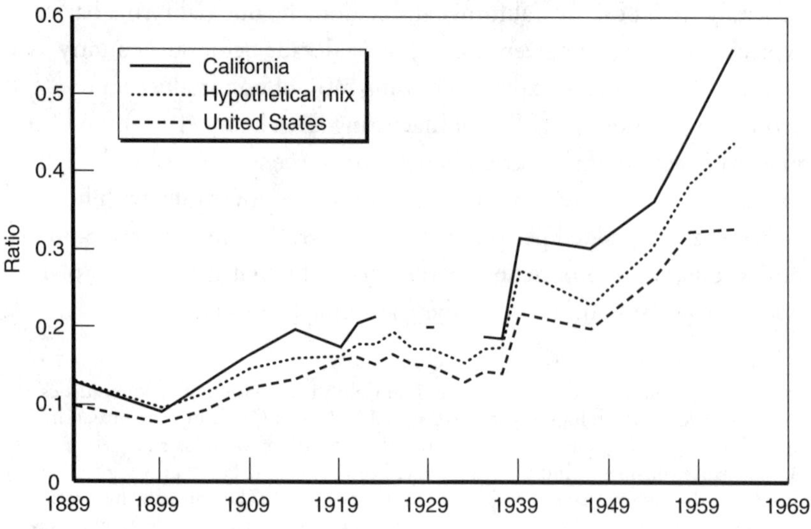

Figure 6.6—Ratios of Nonproduction to Production Workers, 1889–1963

than in the nation as a whole, the average compensation per employee in California was typically much higher than in the country as a whole.

Growing Importance of Headquarters and Auxiliary Functions

In addition to the greater relative prevalence of "white collar" to "blue collar" labor in the state, California has long specialized in central administrative office and auxiliary office (CAAO) activities, which have become increasingly important in the manufacturing economy (Ullman, 1958). This phenomenon was an outgrowth of the state's geographic remoteness and emergence as the center of the western economy. Of course, the San Francisco Bay Area and Los Angeles have never become as dominant a location for corporate headquarters as New York City, and the region has periodically lost control of one or another of its local businesses to national and international interests. But California has continually generated new enterprises to take their places. The state both exerts a disproportionate degree of control over the broader region's economic resources and yet is subject to outside control by larger national and multinational interests.

The Census evidence indicates that the state's concentration of central office and auxiliary functions is not a recent phenomenon. As a result of bookkeeping procedures, the Census Bureau has long kept track of the headquarters location of manufacturing firms. Table 6.2 shows California's share of the nation's CAAO employment and establishments since 1929. The state's share of national CAAO employment has generally risen, climbing from 5.6 percent in 1929 to about 8.1 percent in 1996. The absolute numbers increased almost tenfold over this period, from 11,700 employees to 109,000. The rise has not been continuous—there were reversals during the Great Depression and the mid-1960s. But comparing the state's share of national central administrative office and auxiliary employment to its share of all production workers reveals an even more interesting trend. Before World War II, the state had disproportionately more employees in administrative and auxiliary offices than on production lines. After the war, the relationship reversed. This implies, counter to much of the

Table 6.2

Activity at Central Administrative and Auxiliary Offices in California, 1929–1996

	CAAOs		California's Share of Nation		
			CAAO	CAAO	Production
Year	No. of Establishments	No. Employed	Establishments (%)	Employment (%)	Workers (%)
1929	—	11,688	—	5.6	3.2
1937	213	5,467	6.9	4.5	3.5
1954	—	22,357	—	4.9	6.2
1958	—	34,882	—	5.8	7.2
1963	484	46,920	9.1	6.5	7.4
1967	483	45,900	8.8	5.5	7.5
1972	780	63,100	9.7	6.3	7.5
1977	958	68,600	10.4	6.4	8.3
1982	974	98,500	10.1	7.7	9.8
1987	1,061	93,200	10.7	7.6	10.1
1992	1,064	99,400	9.9	7.9	9.6
1996	—	109,000	—	8.1	9.6

literature that makes World War II the key point when the region grew out of it colonial status, that the headquarters functions were relatively more important before 1940.

Government-Oriented Industries

In yet another pace-setting role, California manufacturers have vigorously pursued the increasing demands of the federal government, especially for defense or aerospace products. The *California Statistical Abstract* has long highlighted the importance of this sector by publishing annual series in its manufacturing chapter on prime defense contract awards to firms in the state and on employment in the broadly defined aerospace (and electronics) sectors.[7] But the best specific evidence on the role of government purchases in manufacturing comes from a set of Census studies initiated with the 1963 "Special Report: Shipments of Defense-Oriented Industries" and continued as its *Current Industrial*

[7]California Department of Finance (various years). For other treatments of this phenomenon, see Clayton (1962, 1967) and Markusen et al. (1991).

Reports, MA-175, until 1983. In its 1963 report, the Census Bureau surveyed 30 industries that "account for the major portion of government procurement" and relied most on demand from the Department of Defense (DoD), the National Aeronautics and Space Administration (NASA), and the Atomic Energy Commission (AEC), predecessor to the Department of Energy (DoE).[8] The published results include the statistics on levels and share of activity (shipments and employment) in each of these industries devoted to meeting military demands and, more important for our purposes, data on the levels and shares by state. Similar surveys conducted by the *Census of Manufactures* in 1987 and 1992 unfortunately lack information about the geographic distribution of federal government purchases.

Table 6.3 summarizes the information from the MA-175 reports by showing California's shares of national employment and value of shipments purchased by the defense agencies. It also displays the shares of these activities in total manufacturing employment and value of shipments for the state and the nation. Over the period of data availability, California constituted a disproportionately large and relatively constant fraction—ranging between 25 and 30 percent—of the nation's manufacturing for the military. In 1963, California led the nation, with government purchases of over $7 billion. This was almost 2.5 times New York's share, the state with the next highest amount of government purchases. California's high share of contracting was largely due to a concentration of its industrial structure on the MA-175 industries and to a lesser extent on the greater orientation of its firms in these industries to filling government demand. In 1963, the value of shipments (government and civilian) of the MA-175 industries constituted about 23 percent of total activity in California compared with 9 percent nationally. Not only was the state's industrial composition weighted toward the MA-175 group, but California

[8]U.S. Bureau of the Census (1965–1966), Vol. I, p. SR2-1. The survey focused on establishments that, either as prime contractors or as subcontractors, had government shipments valued at more than $100,000. See also U.S. Bureau of the Census, "Shipments of Defense-Oriented Industries" (annually, 1965–1977); and "Shipments to Federal Government Agencies" (annually, 1978–1983).

Table 6.3

Defense-Oriented Industries in California Manufacturing, 1963–1992

Year	California's Share of Defense-Oriented Industries: Employment (%)	California's Share of Defense-Oriented Industries: Value of Shipments (%)	Defense-Oriented Industries as a Share of Total Manufacturing — Employment: California (%)	Employment: United States (%)	Value of Shipments: California (%)	Value of Shipments: United States (%)
1963	—	28.9	—	—	19.7	5.8
1965	28.0	26.5	26.1	7.3	19.3	5.8
1966	24.3	25.1	24.7	8.0	20.3	6.6
1967	23.4	23.7	24.9	8.7	20.7	7.4
1968	23.3	22.7	24.8	8.8	19.2	7.2
1969	22.6	22.0	21.1	7.6	16.8	6.4
1970	22.4	22.2	18.3	6.6	15.0	5.6
1971	22.8	23.8	16.4	5.7	13.2	4.5
1972	24.4	26.0	14.4	4.8	11.9	3.8
1973	23.7	25.2	12.4	4.3	10.3	3.4
1974	24.2	26.0	13.8	4.7	10.9	3.6
1975	24.7	25.5	14.9	5.1	11.3	3.8
1976	24.9	26.2	14.6	5.0	11.3	3.7
1977	26.0	27.1	14.6	5.0	11.2	3.7
1978	28.3	26.3	14.2	4.7	10.0	3.4
1979	28.3	25.3	—	4.6	—	3.3
1980	30.4	26.7	—	4.9	—	3.7
1981	28.5	24.1	—	5.3	—	4.0
1982	26.7	25.9	15.8	6.2	12.8	5.0
1983	30.3	28.7	19.1	6.7	16.0	5.6
1987	—	—	—	6.8	—	5.7
1992	—	—	—	4.8	—	4.1

manufacturers in these industries relied more heavily on government demand than their counterparts in the East. For example, the government purchased 87 percent of the value of shipments of California's MA-175 industries in 1963 compared to 67 percent nationally.

The near constancy of California employment and output shares over the 1963–1983 period meant that the MA-175 industries in the state fully enjoyed the stimulus of the spending booms (as in the mid-1960s and the 1980s) and suffered the vicissitudes of the downturns (as in the late 1960s and early 1970s). California's disproportionately high dependence on military demands meant that these fluctuations had greater effects on the state's entire manufacturing sector. This is evident in the changing fractions of total employment and shipments destined for the military. In 1965, for example, over 26 percent of California manufacturing employment produced goods under contracts to the DoD, NASA, or the AEC. This compares to a share of 7.3 percent nationally. The sharp declines in inflation-adjusted defense budgets after 1969 led the California employment share to fall by half to 12.4 percent in 1973. Over this period, the national share fell to 4.3 percent, a more moderate decline in percentage point terms. And with the Carter-Reagan military buildup taking effect in the early 1980s, the California employment share climbed back to 19.1 percent by 1983. The national share stood at 6.7 percent.

Unfortunately, the Census Bureau stopped publishing annual or state-level MA-175 data at this point, making it impossible to trace the further effects of the 1980s buildup or, more important, of the post–Cold War cutbacks of the early 1990s. Judging by the national Census information, it appears that the fraction of 1987 California manufacturing employment devoted to meeting military demands would have been somewhat higher than the 1983 share but lower than the levels prevailing in the mid-1960s.

As an imperfect substitute for contracting data, we can investigate changes in the overall activity levels in the industries that are highly "military-oriented." The category will be defined as those industries for which demand by the DoD, NASA, and DoE accounted for over one-half of shipments. Figure 6.7 shows the shares of the military-oriented industries in total activity in California and the United States from 1939 to 1997. The figure vividly displays California's increasing specialization in military-oriented industries over the first two decades after the end of World War II. In 1963, these industries accounted for almost one-third of California's manufacturing value added and one-quarter of its

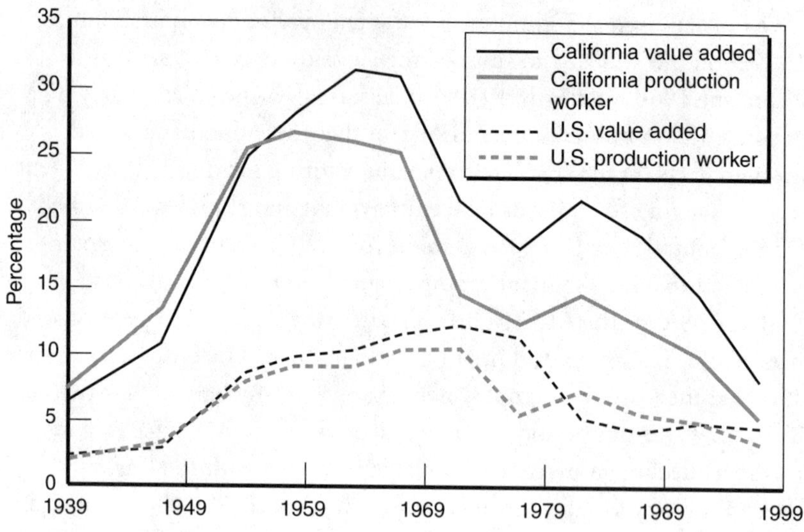

Figure 6.7—Shares of Military-Oriented Industries in Total Manufacturing in California and the United States

production workers. This compared with about one-tenth of manufacturing activity nationally at this date. Note that the California shares peaked in the late 1950s and early 1960s, whereas the U.S. shares continued to grow into the 1970s. Also observe that the buildup in the 1980s appears as a temporary reversal to the secular downtrend rather than a return to the heyday of the 1960s. And the cutbacks of the early 1990s appear in these figures as a continuation of a long-run movement of the California manufacturing sector to diversify away from the military market.

Foreign Markets

Over the last four decades, export markets have emerged as important sources of demand for American manufactured products. This is another crucial arena where California manufacturers were early and more active participants. Table 6.4 provides basic information on the role of foreign markets in manufacturing for California and the

Table 6.4

Export Markets for California and United States Manufacturing, 1960–1997

	California's Share of National Manufacturing		Total "Direct Export" as a Percentage of All Activity				Total "Export-Related" Share of All Activity			
Year	Value of Exports (%)	Value of Shipments (%)	Value of Shipments, California (%)	Value of Shipments, United States (%)	Employment, California (%)	Employment, United States (%)	Value of Shipments, California (%)	Value of Shipments, United States (%)	Employment, California (%)	Employment, United States (%)
1960	9.5	7.9	4.7	3.9	—	—	—	—	—	—
1963	9.5	8.5	4.4	3.9	—	—	—	—	—	—
1966	8.4	8.1	4.1	4.0	—	—	—	—	—	—
1969	9.3	8.4	5.0	4.5	4.6	3.9	—	—	—	—
1971	7.7	8.1	4.7	4.9	4.3	4.1	—	—	—	—
1972	7.7	8.3	4.5	4.8	4.3	4.1	—	—	—	—
1976	9.7	8.6	7.9	7.0	7.7	6.3	—	—	12.3	11.3
1977	11.6	8.9	7.5	6.3	7.7	5.6	—	—	11.7	10.2
1980	10.9	9.5	9.3	8.2	9.2	7.2	13.8	13.5	14.1	12.8
1981	11.4	9.4	9.8	8.0	9.7	7.3	14.2	13.4	14.5	12.8
1983	11.9	10.0	8.2	6.9	7.8	6.0	13.2	12.0	13.5	11.6
1984	11.3	9.8	7.7	6.7	7.3	5.7	13.0	11.9	13.3	11.4
1986	10.8	9.9	7.7	7.1	7.2	5.8	14.0	13.0	14.7	12.6
1987	11.9	10.2	9.1	7.8	7.9	6.3	17.2	15.3	17.0	14.7
1988	12.3	10.1	10.9	9.0	9.7	7.4	16.7	14.7	16.1	13.8
1989	11.7	9.8	12.0	10.3	10.8	8.5	18.3	16.5	17.9	15.5
1990	12.3	10.1	12.4	10.2	11.1	8.6	19.7	17.9	19.2	17.1
1991	12.5	10.1	13.6	11.1	11.8	9.4	21.3	19.4	20.5	18.6
1997	13.6	9.9	16.0	12.0	17.9	13.1	27.6	19.8	28.4	20.4

United States from 1960 to 1997.[9] It compares the growth of California's share of total U.S. manufacturing exports to the state's share of aggregate shipments.[10] Except for a brief period in the early 1970s, California's export share consistently exceeded its output share. Moreover, the growth of the state's export share, which rose from 9.5 percent in 1960 to 13.6 percent in 1997, exceeded the growth of its share of the value of shipments.

The table also shows trends in the export share of manufacturing employment and output in the state and the nation. Two separate series are presented: (1) the jobs and value of shipments in industries "directly" engaged in exporting, and (2) the share of all export-related industries including those producing inputs for the industries directly engaged in exporting. The latter series, created by Census officials using an input-output model, is obviously larger than the former series. All of the series show that exports accounted for a rising share of manufacturing activity. For example, in 1960, direct exports made up about 4.7 percent of the estimated value of shipments in California and 3.9 percent nationally. By 1997, the shares had soared to 16.0 percent and 12.0 percent, respectively. These movements indicate that the growth of export activity has been relatively more important for California manufacturers than for their counterparts in other regions of the country. In the early 1960s, export shares in the state were 0.5 to 1 percentage point above the national average, but by the 1990s, California's shares were consistently 2 to 4 percentage points higher. Clearly, California manufacturers

---

[9]In addition to the *Census of Manufactures,* this analysis relies on U.S. Bureau of the Census. "Survey of the Origin of Exports of Manufactured Products: 1960" (1962); "Survey of the Origin of Exports of Manufactured Products: 1963" (1965); "Survey of the Origin of Exports of Manufactured Products: 1966" (1968); "Survey of the Origin of Exports by Manufacturing Establishments, 1969" (1971); "Survey of the Origin of Exports by Manufacturing Establishments in 1971" (1974); "Survey of the Origin of Exports by Manufacturing Establishments in 1972" (1974)*; 1983 Annual Survey of Manufactures: Origin of Exports of Manufactured Products* (1986); *1984 Annual Survey of Manufactures: Statistics for Industries and Industry Groups* (1986); *Exports from Manufacturing Establishments: 1987* (1991); *Exports from Manufacturing Establishments: 1988 and 1989* (1992); *Exports from Manufacturing Establishments: 1990 and 1991* (1994); and *Exports from Manufacturing Establishments: 1997* (2000).

[10]Because of data limitations, the 1960 and 1963 data use the state's share of value added instead of the share of the value of shipments.

responded earlier and more vigorously to the growing opportunities in export markets. Across many dimensions—in production techniques, business organization, and marketing—California industrial businesses were ahead of the curve over much of the 20th century.

7. Earnings and Productivity in California Manufacturing

California labor has historically been highly productive and earned high returns. This advantageous situation has reflected both the region's scarcity of labor and its population's high levels of skills, educational attainment, and other forms of human capital. Figure 7.1 compares movements in the average earnings for manufacturing employees in California and the country as a whole. It includes annual wages per production worker from 1859 to 1997 and annual salaries per nonproduction employee from 1889 to 1997. Over the late 19th century, California production workers enjoyed a large, but declining, wage differential above the national average. In 1889, the wage premium was 20 percent. With the boom in California manufacturing in the

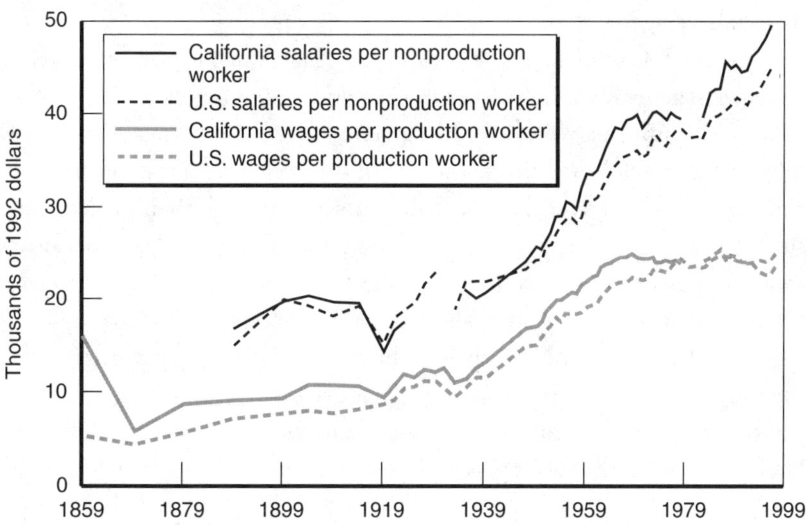

Figure 7.1—Annual Real Earnings per Manufacturing Worker in California and the United States, 1859–1997

1900s, the premium surged. In 1904, wages in the state's manufacturing sector were 34 percent above the national average. This increase was dramatically reversed during World War I, which witnessed the convergence of California's manufacturing wages to the levels prevailing in the eastern industrial belt. Throughout the turbulent interwar period, manufacturing wages in the state remained about 10 to 15 percent above the national average.

Wage movements were less volatile in the period immediately following World War II. From 1947 to 1968, real wage growth was virtually continuous, averaging 1.86 percent per annum in the state and 1.90 percent nationally. This two-decade-long boom was followed by a decade of wage stagnation in the 1970s. The early 1980s witnessed a brief recovery in wage growth, which peaked in 1986. Since that time, however, real manufacturing wages in the state have declined. California lost its long-standing status as a high-wage region, at least compared with other parts of the country. Between 1978 and 1992, annual wages in the state were roughly on a par with those in the country as a whole. In 1997, the most recent year with comparable data, annual wages per production worker in California were roughly 5 percent below the national average.

Although California traditionally was a high-wage region, its salary levels were generally much closer to the national level. Circa 1889, salaries in California were only 13 percent above the national average, a differential roughly two-thirds the size of the wage premium. Salaries generally increased over the next half-century, although their growth was slower and more volatile than that of wages.[1] (This period witnessed important changes such as the increasing feminization of the office staff.) During the interwar period, California briefly became a low-earnings region with salaries 5 percent below the national average. But after 1947, the state salary series again rose slightly above the national series. The gap between the state and nation widened in the 1960s and again in the late 1980s and early 1990s. As with wages, salaries increased consistently

[1]The Census did not collect information regarding nonproduction workers in 1931 and did not publish the data necessary to form state-level aggregates in 1925, 1927, or 1933. The resulting gaps in the series create some minor problems of interpretation.

up to the late 1960s and then stagnated over the 1968–1982 period. In contrast to the wage trends, the salaries in both the state and the nation began a sustained rise in the 1980s. The long-run series reveal that the more rapid growth of California salaries relative to wages was not new, but rather part of a trend dating to the 1920s. One would expect that California's history of income inequality was far different from the national picture.

Trends in the relative labor productivity generally moved in parallel to those of relative earnings (see Table 7.1). In 1859, the value added per production worker in California was 2.7 times higher than the corresponding figure for the country as a whole. This enormous differential undoubtedly reflected the disequilibrium created by the Gold Rush as well as the effects of price level differences. Even after conditions settled down by 1879, however, the measured "productivity" differential stood at roughly 50 percent. As with wages, the differential declined over the late 19th century, reopened between 1899 and 1914, and then fell again during World War I. Over the interwar years, the productivity ratio typically fluctuated in a range between 1.22 and 1.32. As with the earlier wartime period, the California-U.S. productivity gap closed during World War II. By 1947, for example, the ratio stood at 1.21, down from 1.32 in 1939. Since 1947, the productivity differential has moved in several waves—down over the 1947–1955 period, up from 1955 to 1964, and sharply down in the 1970s. By 1997, the ratio hovered around 1.10. Another pertinent way to evaluate the productivity differential is to explain trends in output per hour of labor of production workers. The series comparing value added per hour of labor by production workers in California with the country as a whole over the post–World War II period reveal largely similar patterns to those for annual output per production worker. Finally, the table also presents a series on relative value added per employee (including both production and nonproduction workers.) The differential in the total employee ratio is smaller than in the production worker series, reflecting the state's higher ratio of nonproduction to production workers. In 1939, for example, the state value added per employee was 22 percent above the national series and the value added per production worker series was 32 percent higher. By 1997, the ratios had declined to 0 percent and 10

Table 7.1

Ratio of California/United States Real Value Added per Unit of Labor, 1859–1997

Year	Production Worker	Production Worker-Hour	Employee
1859	2.71	—	—
1869	1.42	—	—
1879	1.51	—	—
1889	1.31	—	1.28
1899	1.21	—	1.20
1904	1.33	—	1.29
1909	1.41	—	1.35
1914	1.39	—	1.32
1919	1.15	—	1.14
1921	1.25	—	1.21
1923	1.28	—	1.22
1925	1.19	—	—
1927	1.21	—	—
1929	1.24	—	1.19
1931	1.32	—	—
1933	1.36	—	—
1935	1.31	—	1.26
1937	1.23	—	1.19
1939	1.32	—	1.22
1947	1.21	1.22	1.16
1954	1.17	1.17	1.12
1958	1.19	1.17	1.12
1963	1.22	1.22	1.09
1967	1.19	1.19	1.09
1972	1.17	1.19	1.08
1977	1.12	1.12	1.05
1982	1.17	1.16	1.09
1987	1.09	1.13	1.02
1992	1.15	1.15	1.03
1997	1.10	—	1.00

percent, respectively. According to the value added per total employee measure, labor productivity in California manufacturing has completely converged to the national average.

To summarize, California has historically been a high-wage, high-productivity region. But in recent years, wages of production workers in the state have stagnated, leading the wages of other regions to catch up with and move slightly ahead of those in California. By way of contrast, the earnings of nonproduction workers have continued to rise. Although California manufacturing is no longer characterized by high wages, it continues to generate relatively high labor returns.

8. California's Small-Plant Economy

Over most of the 20th century, large multi-division corporations were viewed as the model for modern business organization (Chandler, 1977). Small businesses were seen as socially desirable but something close to an endangered species. Such attitudes changed after the mid-1970s with the growing dissatisfaction with bureaucracy, the slowdown in the growth of the Fortune 500 firms, and the fantastic success of startups in the high-technology fields. Discussions highlighting the vital economic role of the entrepreneur moved from the academic writings of Joseph Schumpeter to the front pages of the business press. Government agencies and chambers of commerce began charting trends in small business formation because they were viewed as the economy's most powerful engines of job creation. Also contributing to this change in public attitudes was the downsizing of the American corporate sector in the early 1990s and the emphasis on flexible networks of small, innovative firms (Saxenian, 1994).

The statistics on California manufacturing show that the smaller-scale business has always been relatively more important in the state than in the country as a whole. Over most of the 20th century, the size of California manufacturing plants, as measured by production workers per establishment, was consistently below that of their counterparts outside the state (see Figure 8.1).[1] In 1869, there were 7.9 production workers per establishment in California compared with 9.4 nationwide. Accompanying the rise of big business nationally, the number of production workers at the typical U.S. manufacturing establishment in 1899 climbed to 17.7 (if factories performing custom operations are

[1] Patterns for total employment per establishment and value added per establishment were broadly similar.

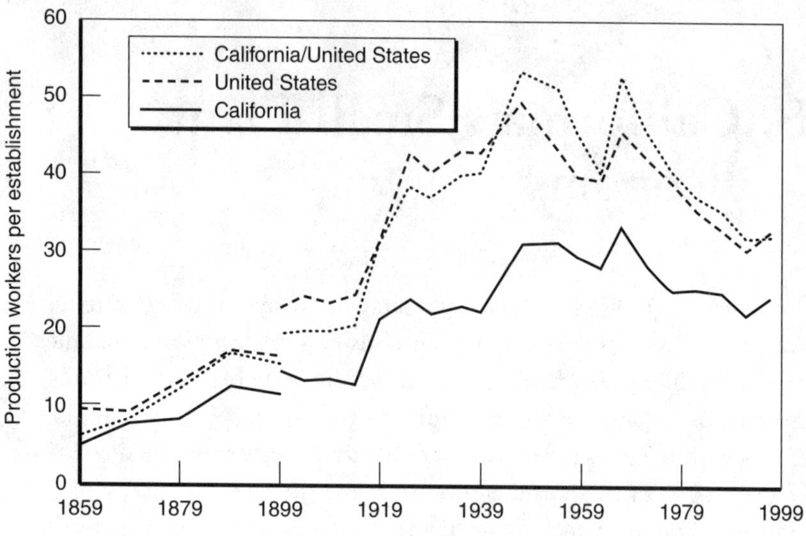

Figure 8.1—Establishment Size in California and the United States, 1859–1997

included and 23 if they are excluded). Over the early 20th century, average plant size continued to increase nationally, with the number of production workers per establishment reaching 31.3 by 1919 and 42.9 by 1939.

The growth of the size of manufacturing plants in California did not keep pace. In 1899, the number of production workers at the typical California manufacturing establishment was 11.7 (if factories performing custom operations are included and 14.7 if they are not). Focusing on the numbers consistent with the time series for the late nineteenth century, plant size in California was about 70 percent of the national figure and was roughly on par with what prevailed nationally two or three decades earlier. Even as the state's manufacturing sector began to expand more rapidly after 1899, employment per plant in California did not grow as fast as it did nationally. In 1939, for example, the average California establishment had 23.5 production workers, 61 percent more than in 1899, but about half the national average.

Between 1939 and 1954, the size of California manufacturing plants began to converge toward the national average. By the end of that period, the typical California plant had 31.6 production workers, one-third more than in 1939 and nearly 73 percent of the national average in 1954. The change from 1947 to 1954 was almost as great as that between 1939 and 1947, indicating that the shift was not purely the result of wartime changes. Over the last quarter century, manufacturing plants have tended to downsize. Between 1967 and 1997, the average size of plants in California fell from 32.7 production workers to 24.1, which was close to the state's average during the interwar period. The national trend of worker per establishment was roughly similar, and the ratio between the state and national plant sizes remained stable at about 70 percent.

Before 1939, the small size of California plants had been a major cause of local concern, especially regarding whether the state's manufacturing sector could fully exploit the available economies of scale to compete effectively with larger eastern plants. In the post-1939 period, the rise of average plant sizes in California and the closing gap between the state and national figures were taken as signs that California's industrial sector was maturing, overcoming some of the previous limitations of its development. In light of these previous concerns, it is curious that the decline in plant size in California over the last quarter century has not led to serious hand-wringing. If anything, that decline has drawn applause because the state is now viewed as an incubator for small, dynamic firms.

An important part of the growth in plant size over the 1939–1967 period was due to shifts in the state's industrial distribution. Figure 8.1 also compares the actual state and national average plant sizes with a hypothetical series generated by combining the national plant sizes for individual industries with the California industrial distribution.[2] This analysis indicates that over most of the late 19th and early 20th centuries, there was a small-plant bias in California's industrial structure. For example, in 1899 when the average plant nationally had 22 production workers, the hypothetical combination of California distribution and the

[2]If the hypothetical series is below the actual national series, then California's structure was biased toward small-plant industries; if the hypothetical series is above, the state is biased toward industries that had large plants nationally.

national plant sizes yielded a figure about 15 percent lower. Over the first half of the 20th century, the distribution gradually shifted in favor of industries characterized by larger plants. By 1954, the hypothetical California series stood above the actual national series, indicating a bias in the state's structure toward large-plant industries. This shift reflects, in part, the increasing importance of the aerospace complex in the state, which was characterized in the post–World War II period by huge plants nationally. (In recent years, there has been little bias in the structure in either direction.) Note that in the typical industry, manufacturing plants remained smaller in California than in the nation as a whole over the entire post-1939 period.

The plant size issue has received some criticism in the social science literature (Granovetter, 1984). Because the denominator treats all plants equally, the measure weights small plants disproportionately relative to their importance in the economy and potentially creates misleading impressions about changing scale. For example, the literature has noted that the national rise in average size between 1919 and 1967 was primarily due to a decrease in the number of establishments with five or fewer workers. These small operations made up less than 4 percent of employment even at the beginning of the period. There was little change in the size distribution of plants having above 20 workers, which accounted for most employment. Hence, it would be a mistake to think that the typical production worker in 1967 labored in a plant that was vastly larger than his or her counterpart in 1919. The question arises whether the differences between the state and the nation and their changes over time reflect genuine across-the-board differences in the size distribution.

Between 1909 and 1939, the Census presented the size distribution of establishments categorized by the number of production workers per establishment. Figure 8.2 compares the size distribution in California and the United States as a whole in 1909, 1919, 1929, and 1939. For each geographic unit, the figure shows the cumulative proportion of establishments and production workers in each size category. Establishments in the largest categories were much scarcer in California than they were nationwide. For example, in 1909, the establishments with more than 500 wage-earners accounted for 13 percent of

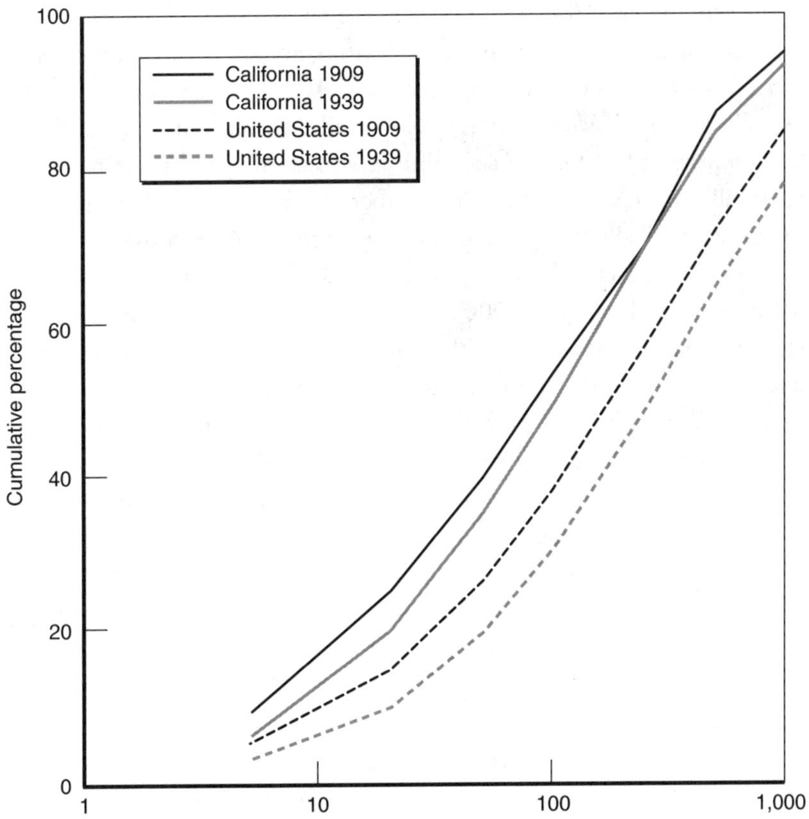

Figure 8.2—Distribution of Production Workers by Establishment Size

production workers in California compared to 28 percent in the United States. These movements indicate that focusing on the average plant size is not likely to result in a seriously misleading picture of the state's development.

Why was there a greater prevalence of small plants in California? One possible explanation is that the pattern reflects the state's more active entrepreneurial community, its more vigorous venture capital markets, and generally healthier environment for small businesses. Another explanation focuses on constraints rather than opportunities. It argues that the high transportation barriers created a small home market for the region's manufacturers and prevented its plants from realizing all of the economies of scale that plants in the East enjoyed. This explanation was a standard mid-20th century account of why California's manufacturing was relatively undeveloped.

The *Census of Manufactures* itself provides evidence in favor of the hypothesis that California's economic environment acted as an incubator for small ventures. Although the number of production workers per capita was typically below the national average, since 1899 the number of manufacturing establishments per capita in California has been above the national ratio. For every 1,000 residents in 1899, for example, there were 3.4 manufacturing establishments in California compared with 2.7 nationally. Had the California ratio equaled the national one, and holding the number of production workers constant, the average size of California plants would have been 83 percent of the national average instead of 66 percent. Over the 1899–1939 period, the ratio of establishments per 1,000 residents fell both in the state and in the nation as the population increased faster than the number of establishments. (The Great Depression disproportionately eliminated smaller firms.) But the differential between California and the United States as a whole remained. By 1939, the ratio of plants per 1,000 people was 1.4 nationally and 1.7 in the state.

Yet a strong economic argument can be made that it is more relevant to compare California's share of national establishments with its share of the total economy, that is, with its income share rather than its population share. By this measure, California's emergence as an incubator for manufacturing ventures dates to the 1970s rather than to the 1900s. This argument is consistent with many traditional accounts that concluded that the small size of the region's markets seriously constrained the state's manufacturing sector. For example, Gordon (1954) called the inadequate size of the western market "the most important factor that has hampered the growth of manufacturing in the state." Kidner (1946, p. 28) argued that the "absolute size of the population of California did not constitute a market large enough to warrant large-scale manufacturing until after 1900." Because of the small market, California firms could not produce "on a sufficiently large scale" to offset the competitive advantages of eastern producing centers.[3]

Another piece of evidence in favor of this position is that establishment sizes in California were typically closer to the national average in industries for which the state was an exporter and thus not limited to the region's market—that is, in food processing and in transportation after 1939. We also know that as the size of the region's market grew, the state's industrial structure shifted from a bias toward industries with small size nationally to those with large size nationally.

Perhaps it would be better to view the two hypotheses as complementary, rather than conflicting, explanations. One might speculate that given the constraints the region suffered in moving into larger-scale activities, California developed and retained institutions and markets suited to the encouragement of smaller-scale enterprise. When the direction of technical change shifted during the last decades of the 20th century, these capabilities redounded to the region's advantage in becoming an incubator for small, innovative firms. Thus, the constraints in one period created opportunities in the next.

---

[3]See also Niklason (1930, p. 404), and Grether et al. (1946), especially Ch. IV.

9. Putting the Pieces Together

Although it is impossible to provide a simple and complete explanation of a historical process as complex as the rise of California to national industrial leadership, the evolution of California's manufacturing sector can be interpreted usefully as the result of six broad, long-run processes.[1]

Resource-Based Growth

The first process is industrial growth based on the discovery and exploitation of California's distinctive natural resource base. The Gold Rush expansion is the most obvious and widely known example of resource-based development in the state. The emergence of the flour milling industry during the California's wheat farming period from 1860 to 1890 represents a second case. Still others are the growth of canneries, wineries, and sugar beet processing plants during the late 19th century when the state's agricultural sector shifted from extensive to intensive crops. The discovery of vast oil reserves in the 1900s and 1920s provided the resource base for the state's large petroleum refining industry and stimulated industrialization in general by significantly lowering energy costs. The case of oil is especially interesting because it is an example of resource-based growth based on import substitution. Replacing imported coal and kerosene and producing gasoline for the western regional market drove the petroleum industry's early expansion; it was not substantially export-oriented until the 1920s. This example also belies the commonplace that the Golden State simply reaped the benefits of a rich and varied natural resource base. The petroleum discoveries assumed the importance they did because California almost completely

[1]Each process operated with special intensity during particular periods, but inserting discrete breakpoints in the historical narrative is ultimately unhelpful, for even if a break in trend could be statistically observed in the time-series data, it may not represent the date when the fundamental change began.

lacked coal. The state's natural environment, although distinctive, was not entirely conducive to its industrial development.

Product Market Integration

The second force conditioning California's industrialization was its growing integration with the national and global economy as transportation costs fell. This integration opened new export markets and increased competition from imports. Over the late 19th century, real transcontinental railroad rates declined sharply so that the rates in 1900 were only one-fourth their 1870 level. After 1900, the downward trend had largely ended and transcontinental rates were little changed in real terms until the 1950s.[2]

California adjusted to the falling transportation rates in several ways. A wide range of industries, including textile mill products, boots and shoes, and tobacco products, contracted in relative, and in some cases, absolute terms. Without very high transport barriers, the California firms could not compete in producing the light consumer goods that were marketed on a national basis. As we shall see below, other factors such as the cutoff in Chinese immigration also contributed to the decline of these industries (Johnson, 1985).

Many of California's resource-processing industries, including its canning and winemaking sectors, greatly expanded to serve the widening export market. Employment in export-oriented agricultural processing industry grew from fewer than 1,000 production workers in 1869 to nearly 10,000 in 1899. The "tyranny of distance" did not completely lose its influence. Transportation barriers help explain why the horticultural crops of the state were so extensively processed: Producers needed to reduce perishability in transit, eliminate waste, and raise their products' value-to-weight ratio. More generally, the relatively higher cost of reaching external markets accounts for the general emphasis of

---

[2]These rates were not perfectly stable. Real rates fell during World Wars I and II and rose during deflationary episodes. Given government control of the railroads during the wartime period, the lower rates probably did not translate into lower actual transportation barriers in shipping between the East and West. (U.S. Interstate Commerce Commission, 1903; various years, 1887–1960.)

California exporters on "adding value" by imbuing their goods with higher quality or a more attractive image (Cleland and Hardy, 1929).

Over the mid-20th century, the state developed a specialization in transportation equipment, specifically aircraft, to help overcome these barriers. Aircraft are both a means of transportation and a high-value-to-weight product. The industry is normally oriented to export and military markets. But it is worth noting that the Pacific Coast industry achieved its aircraft leadership based on its success with planes designed and built to serve the commercial aviation market in the West. This success itself was based on technological innovations of a small cluster of aviation engineers and entrepreneurs. The region's firms acted as the vanguard of the "airframe revolution" of the early 1930s when they designed and produced the first modern airliners as exemplified by Boeing's 247 and Douglas's DC-2 and DC-3. These streamlined, all-metal, cantilevered monoplanes were built at the request and to the specifications of the transcontinental airlines—United, TWA, and American—specifically for the purpose of serving their routes between the Pacific Coast and the East. Riding on these achievements, California became home to over half the nation's aircraft industry on the eve of World War II. The development of this industry during and after the war helped the state's manufacturers pursue external markets in the postwar period.

Factor Market Integration

At the beginning of the American period, California had high relative wages and even higher returns to capital. As transportation costs fell and outsiders became increasingly familiar with the region, the state's factor markets became better integrated into the world economy. We tend to view this as an inevitable or natural process, but it is by no means certain where the strongest cumulative flows of information, labor, and capital will develop. Policy choices and accidents help determine which of many alternative paths is followed.

During the mid-19th century, for example, California was beginning to develop significant population inflows from Asia, but labor market integration was cut short by anti-Chinese agitation and the 1882 Immigration Restriction Act. This had important consequences for California manufacturing. The policy shifts in combination with the influx of eastern-made goods undermined the competitive position of California manufactures in such low-wage industries as cigars, boots and shoes, and apparel. Instead, California forged stronger labor market connections with the midwestern region of the United States, an area characterized by high wages and high levels of investment in human capital. The resulting migration flows helped the state develop an educated labor force at lower cost. As a consequence, the state's manufacturing development over much of the 20th century was channeled along a high-wage path, based on industries using high ratios of nonproduction to production workers and advanced technologies.

Policy changes in the 1960s, including the end of the Bracero program in 1964 and the 1965 revisions of the 1924 Immigration Quota Act, again shifted the migration pathways. Since that time, California has attracted large population inflows from Latin America and East Asia. These flows have substantially enlarged the pool of less-skilled labor and fundamentally changed California's status as a labor-scarce region. These changes have resulted in complementary shifts in the state's comparative advantage, leading a cluster of low-wage manufacturing sectors (including apparel and leather goods) to expand rapidly since 1970.

Demand-side developments complemented these changes in supply. The apparel industry, which had a long history in the region, survived competition from lower-wage regions by focusing on a distinctive set of specialties. These included more casual clothing linked to the western lifestyle. One example is Levi's jeans, which gained worldwide popularity in the post–World War II era. A second area of specialization was high fashion, where development was tied to the rise of Hollywood as a global center of the glamour industry (McWilliams, 1949, p. 218–220).

Growth Based on the Home Market Size

Historically, a large segment of California's manufacturing industries was local or regional in scope. Given the state's high transportation costs, many manufacturing products had limited import or export possibilities. Included in this group were printing products such as newspapers, many food products such as bread, dairy, and other perishables, most building materials, and many of the metal and machinery products of firms specializing in custom work or repair. For printing and nonexport food products, the home demand depended primarily on the *size* of the region's population and income, whereas for building materials, demand depended primarily on the *growth* of the region's population and income. Between 1890 and 1960, the state's population growth displayed a saw-tooth pattern of alternating decades of rapid and slow expansion. For most of its history, printing employment closely mimics population growth. Forest products and stone/clay/glass experienced a much more exaggerated saw-tooth pattern. Growth in California was highly positive during the decades of rapid population growth—the 1900s, 1920s, and 1940s—and was low or negative during the decades of slow population expansion—the 1890s, 1910s, and 1930s.

Much of the machinery and metal trades grew up to serve the region's producers in agriculture, mining, and manufacture itself. By the late 19th century, the state supported a flourishing foundry and machine shop industry as evidenced by the size of the fabricated metal group. Plants in these industries typically operated on a smaller scale than their counterparts in the East and were oriented to a custom and repair trade. In many cases, especially in the mining and agricultural machinery wings, the local firms perfected and produced equipment specially designed for western economic and environmental conditions. The combined grain harvester and the track-laying tractors, innovative technologies developed by the farm machinery firms centered in Stockton, were two of the most notable machines initially designed for western conditions that eventually captured world markets. The move to wider markets was associated with the relocation of the leading California farm machinery firms to the industrial belt (Olmstead and Rhode, 1988). These industries expanded locally in response to growth in activity in the region's manufacturing, farming, and mining sectors.

Scale-Dependent Import-Substitution

The fifth process is a hybrid of the second and the fourth resulting from the presence of scale economies internal to the plant or firm. For a range of products, especially in the mass-production industries, the West could compete in its own market with eastern imports, but only if the scale of operations was sufficiently large. According to Gordon and Kidner, California's market size in the late 19th century was often inadequate to support modern large-scale production. Indeed, they argue that the rapidly increasing scale of efficient plants over this period undercut the state's industrialization efforts.[3] The more vigorous market expansion of the 20th century pushed a larger number of firms and industries across the scale threshold of profitable production in the region. The volume and variety of goods produced locally increased in a process of scale-dependent import-substitution.

The experiences of California's automobile and tire industries offer valuable, concrete examples illustrating this process. Modern automobile production in California began with Henry Ford's establishment of assembly plants in San Francisco and Los Angeles in 1914. In the next two decades, Chevrolet (1916), Chrysler (1932), Buick-Oldsmobile-Pontiac (1936) and a handful of other minor producers set up branch plants in California. The inflow continued after 1945. By 1950, every important automaker, except Cadillac and Hudson, has established plants in California and Los Angeles assembled more different makes of automobiles than Detroit. The western tire industry reached nearly full scale in even less time than the auto industry. In 1920, Goodyear Tire opened a large tire plant employing modern straight-line production techniques in Los Angeles, thus becoming the first firm of the industry's Big Four to open a plant on the West Coast. Within a decade, Goodrich, Firestone, and U.S. Rubber had also built plants in Southern California. By the mid-1930s, Los Angeles could boast a tire industry second only to Akron and an auto assembly industry second only to Detroit.[4] This form of growth was not limited to large-scale branch plants of national firms. In many markets with high degrees of product differentiation, a wider variety of smaller-scale producers could enter as the region's economy increased in size. The process appears especially

---

[3]Gordon (1954, pp. 36, 56–57, 63, 70); Kidner (1946, p. 28).

[4]Most of the rubber tire, auto assembly, and integrated steel plants in the state closed in the 1970s and 1980s in response to further reductions in transportation costs, increased international competition, and the national process of industrial restructuring. In a sense, the second process—product market integration—finally undermined the operation of the fifth process.

important in explaining California's success in creating sufficient jobs for its vastly enlarged labor force in the conversion period after World War II.

External Economics of Scale

The sixth growth process is industrial expansion resulting from external economies of scale, either in inputs as the economy grew or through localized technological spillovers. The most notable examples of this process were among the state's cluster of high-technology firms, especially in the aircraft and electronic industries. As noted above, the aircraft industry is one of the few manufacturing activities that could be called a leading sector. Airplane building began in California just six years after the Wright Brothers historic flight at Kitty Hawk, North Carolina. In 1909, Glenn Martin started his long career in aviation in the state of California, making and flying an aircraft at Santa Ana. For the next eight years, Martin was at the center of the West's small, tight-knit aviation community. It was Martin who taught Seattle lumber magnate William Boeing to fly in 1915. It was Martin who first attracted aeronautical engineer Donald Douglas to design airplanes at his Los Angeles plant in 1915–1916. But during World War I, Martin relocated his aircraft operations to the East because Southern California's input markets were not yet deep and diverse enough to support the airframe industry on a large scale.

The Pacific Coast industry essentially began anew in the early 1920s. The modern industry is usually dated from Donald Douglas's return to California in 1920 to establish a small local shop. The Lockheed brothers and John Northrop also established cutting-edge aircraft operations in Southern California. Another important development in the late 1920s was the emergence of Caltech as a center for aeronautical research with close ties to the community of engineers and entrepreneurs in the local aircraft industry. By the early 1930s, production capabilities and infrastructure were in place for the industry to lead in the "airframe revolution." Following this success, eastern firms relocated to the West to be close to the evolving new technologies, and startups tapping the local pools of engineering talent and venture capital added to the boom.

The evolution of the state's electronics industry paralleled that of the aircraft industry in many ways. Both industries manufactured products that withstood the "tyranny of distance" well. As in the aircraft industry, the early pioneers in the state's electronics industries, such as Lee Deforest and Philo Farnsworth, found it difficult to achieve success in the immature California economy of the early 20th century. However, the electronics industry in California benefited from technological and demand spillovers from the development of the state as an aviation center. The state's aircraft producers as well as its entertainment firms were important early customers for the electronics sector. Finally, as in the aircraft industry, the success of California's electronics industry was the outgrowth of the innovative efforts of small-scale startups (Hewlett-Packard, Varian, and others), the formation of a collaborative network of entrepreneurs and engineers, and close connections with leading research universities, primarily Stanford.

It is important to stress the key historical role of the overall expansion of the regional market, on both the input and output sides. Despite a long tradition of innovation and risk-taking in California dating back to the Gold Rush, the state would not have emerged as a global center of high technology in the second half of the 20th century without the fundamental contributions of the multifaceted forces of growth that stimulated the rapid expansion of what had been a small, remote outpost in the mid-19th century.

10. The Evolving Role of Manufacturing in the California Economy

Industrialization began later in California than in the nation as a whole. And the state's path of manufacturing development did not follow older examples based on employing low-wage labor in large factories or using coal for smelting and steam power generation. The trend toward deindustrialization also began later in the state than in the nation, because of the state's specialization in high-technology and defense industries as well as its movement into apparel production. As this chapter documents, using labor force and earnings data, the role of manufacturing in the California economy has evolved dramatically.

This analysis shows that during the key period during the middle of the 20th century when California emerged as a global center of economic activity, the manufacturing sector was one of the key *driving forces* for the region's economic growth. But more generally, the sector's expansion was one part of a balanced growth process in an economy that was "running on all cylinders."

Another important finding is that California has always been less dependent on manufacturing activity than the nation as a whole. I will argue that this should be taken as a sign of strength rather than of weakness because the state has long been characterized by a relatively advanced economic structure with a high-productivity agricultural sector and a large, diverse service sector. Unlike less fortunate areas, California did not require manufacturing to serve as the sole "engine of growth" to drag other lagging economic sectors forward. Of course, this finding in no way implies that industrialization in California was unimportant. In the absence of industrial development, the state would have had to import a larger fraction of manufactured goods, especially automobiles and other machinery. Moreover, the state's economy would have been able to support a much smaller population at prevailing income levels and a much less diverse range of economic activities.

Tables 10.1 and 10.2 present data by decade from the *Census of Occupations* on the distribution of California's labor force across major industrial categories between 1870 and 1990.[1] It also compares California's occupational structure with that of the country as a whole. (A break in the series, created by a change in the Census definitions between 1930 and 1940, necessitates presenting the data in two parts. The second table, covering 1930–1990, presents shares of the "civilian" labor force.) These labor force data reveal that California's occupational distribution was always fairly balanced across sectors and, over time, became more similar to the national structure. For example, in 1870, it would have been necessary to shift about one-quarter of California's labor force across the 11 sectors to match the national distribution. By 1990, the fraction fell below 4 percent, a remarkably low number. The division of employment across the three major occupational categories—primary, secondary, and tertiary—was almost identical in California and the United States as a whole. This reflects the common observation that in many ways California was like a nation unto itself.

During the second half of the 19th century, the manufacturing sector's share of the total labor force was consistently lower in California than it was nationally. For example, in 1870, 14.2 percent of the state's labor force was employed in "manufacturing and mechanical occupations" compared with 16.2 percent nationwide. And in 1900, these workers constituted 17.7 percent of the labor force in California and 19.4 percent nationally. The manufacturing sector's share in California generally increased over the first half of the 20th century, rising to about 25.3 percent in 1960. Growth was especially robust during the 1910s, 1940s, and 1950s.[2]

---

[1]Gordon (1954, pp. 170–171); U.S. Department of Commerce (1975); U.S. Bureau of the Census, *1980 Census of Population* (1983, 1984); U.S. Bureau of the Census, *1990 Census of Population, Social and Economic Characteristics: California* (1993a, 1993b).

[2]California's construction sector was also relatively larger than it was nationally. This, of course, is what one would expect in a region experiencing rapid increases in

Table 10.1

Composition of California's Labor Force, 1870–1930

Sector	1870	1880	1890	1900	1910	1920	1930
				California			
Agriculture	29.3	28.6	29.0	25.0	17.9	17.3	13.3
Forestry and fishing	1.7	2.1	1.7	1.4	1.2	0.8	0.5
Mining	15.5	10.3	4.6	4.7	2.8	1.7	1.6
Construction	5.9	5.3	6.4	6.1	9.0	7.4	7.5
Manufacturing and mechanical	14.2	17.5	16.7	17.7	17.9	21.0	18.0
Transportation and communication	6.7	6.9	7.4	8.4	9.5	8.3	8.0
Trade	9.2	10.9	10.8	13.7	13.9	13.8	17.5
Public service	1.3	1.3	1.3	1.6	2.2	3.0	2.4
Professional service	2.9	3.7	5.1	6.2	6.3	7.7	9.4
Domestic and personal service	11.7	11.7	14.0	12.7	12.8	10.2	11.8
Clerical workers	0.8	1.6	2.9	3.0	6.2	8.8	10.1
Commodity-producing	66.5	63.9	58.5	54.8	48.8	48.1	40.9
Primary	46.4	41.1	35.3	31.1	22.0	19.7	15.4
Secondary	20.1	22.8	23.2	23.8	26.8	28.4	25.5
Tertiary	32.6	36.1	41.5	45.5	51.0	51.8	59.1
				California/United States Ratio			
Agriculture	0.56	0.58	0.70	0.67	0.58	0.64	0.62
Forestry and fishing	4.45	4.01	2.21	2.03	1.84	1.25	1.02
Mining	10.52	6.08	2.40	1.94	1.10	0.64	0.79
Construction	1.15	1.11	1.11	1.11	1.46	1.42	1.23
Manufacturing and mechanical	0.88	0.99	0.91	0.91	0.80	0.84	0.79
Transportation and communication	1.50	1.41	1.16	1.23	1.35	1.14	1.01
Trade	1.49	1.47	1.31	1.38	1.64	1.38	1.40
Public service	1.74	1.67	1.46	1.60	1.79	1.68	1.39
Professional service	1.10	1.17	1.33	1.51	1.42	1.52	1.41
Domestic and personal service	1.19	1.35	1.34	1.27	1.27	1.28	1.16
Clerical workers	1.26	1.69	1.36	1.13	1.34	1.19	1.23
Commodity-producing	0.88	0.86	0.86	0.84	0.78	0.80	0.77
Primary	0.86	0.80	0.80	0.76	0.64	0.65	0.64
Secondary	0.94	1.01	0.96	0.95	0.94	0.94	0.88
Tertiary	1.33	1.40	1.30	1.32	1.42	1.31	1.25

population. Growth in this sector was especially rapid in the 1900–1910 decade, when population expansion and the recovery from the 1906 earthquake pushed up building activity, and in the 1940–1950 decade. The construction sector expanded slowly in the 1910s and 1930s, when population growth sagged.

Table 10.2

Composition of California's Labor Force, 1930–1990

Sector	1930	1940	1950	1960	1970	1980	1990
California							
Agriculture	14.1	10.9	7.5	4.8	3.0	2.9	3.0
Forestry and fishing	0.6	0.3	0.3	0.2	0.2	0.2	0.1
Mining	2.1	1.9	0.8	0.5	0.5	0.4	0.3
Construction	6.7	6.2	7.8	6.7	5.4	5.7	6.8
Manufacturing	17.4	17.0	19.8	25.3	21.6	20.3	16.9
Transportation, communication and public utility	8.3	8.2	8.4	7.1	7.1	7.1	6.7
Trade	17.8	22.6	22.6	19.7	21.1	20.9	20.9
Finance, insurance, and real estate	5.3	4.9	4.6	5.3	5.9	7.1	7.6
Professional service	8.4	8.8	10.1	13.0	18.5	20.0	21.9
Other service	16.0	15.1	11.7	11.2	10.4	10.3	11.4
Public administration	3.2	4.0	6.4	6.3	6.4	5.1	4.4
Commodity-producing	40.9	36.3	36.1	37.4	30.6	29.4	27.1
Primary	16.8	13.1	8.6	5.4	3.6	3.5	3.4
Secondary	24.1	23.3	27.6	31.9	27.0	26.0	23.7
Tertiary	59.1	63.7	63.9	62.6	69.4	70.6	72.9
California/United States Ratio							
Agriculture	0.62	0.57	0.60	0.70	0.83	0.90	1.17
Forestry and fishing	1.04	1.32	1.36	1.13	1.20	1.24	0.84
Mining	0.90	0.90	0.48	0.45	0.55	0.40	0.45
Construction	1.17	1.33	1.24	1.07	0.90	0.80	1.10
Manufacturing	0.76	0.72	0.76	0.90	0.83	0.96	0.95
Transportation, communication and public utility	1.09	1.16	1.06	1.00	1.05	1.08	0.94
Trade	1.33	1.32	1.19	1.04	1.05	1.02	0.98
Finance, insurance, and real estate	1.73	1.48	1.35	1.22	1.18	1.18	1.10
Professional service	1.30	1.18	1.18	1.07	1.06	1.00	0.94
Other service	1.29	1.26	1.22	1.14	1.19	1.15	1.21
Public administration	1.23	1.22	1.42	1.24	1.18	0.95	0.93
Commodity-producing	0.75	0.73	0.77	0.88	0.84	0.91	0.99
Primary	0.66	0.61	0.59	0.67	0.79	0.80	1.02
Secondary	0.84	0.82	0.85	0.93	0.85	0.92	0.99
Tertiary	1.29	1.27	1.20	1.09	1.09	1.05	1.00

NOTE: For 1930–1990, series cover civilian labor force only.

The labor force data reinforce the perception that the manufacturing sector was the driving force for growth over the crucial 1940–1960 period, when the total civilian labor force in California more than doubled in size, increasing from about 2,480,000 to over 5,760,000. The manufacturing labor force increased from about 420,000 to nearly 1,460,000 and accounted for a little less than one-third of aggregate growth. Indeed, the manufacturing sector's contribution to the aggregate expansion was almost twice what would be expected under a balanced growth process.

After 1960, however, the manufacturing share of the labor force fell both in the state and the country as a whole. Up to 1990, the sector's relative decline was more rapid for the United States than for California. In 1990, the manufacturing share of the state's labor force stood at 16.9 percent, 8.4 percentage points below the 1960 peak and roughly on a par with the 1940 share. For the United States, the sector's labor force share in 1990 was 17.7 percent, which was 10.5 percentage points below the 1960 peak and 6.1 percentage points below the 1940 level. In 1980 and 1990, the manufacturing share in California came as close to parity with the national share as it had been for one hundred years.

The smaller historic role of manufacturing in California does not appear to be due to a specialization in the primary sector (agriculture, forestry, fisheries, and mining). Contrary to the conventional wisdom about California, extractive activities traditionally constituted a smaller fraction of total employment in the state than they did nationally.[3] In 1900, for example, agriculture accounted for about 25 percent of the

---

[3]California was distinctive because the gap between output per worker in the extractive sector and the nonextractive sector—the so-called development gap—was much smaller in California than in the country as a whole. For example, in 1880, the annual output of the average member of the agricultural labor force was about 44 percent that of the average member of the nonagricultural labor force in the United States as a whole, whereas the ratio was above 81 percent in California. A similar picture prevailed in 1960, when the national ratio stood at 48 percent and the state ratio at 80 percent. Over the late 19th and early 20th centuries, the development gap in both areas tended to widen and then contract, but it remained consistently less than one-half as large in California as in the nation. The conventional explanations of the gap emphasize the lower number of hours worked in agriculture because of seasonal underemployment, the high rates of natural increase of the farm population, and the barriers to labor mobility out of activities experiencing relatively slow rates of demand and productivity growth. Historically, these forces have exerted less weight in California's diverse, expanding, and progressive agricultural sector than elsewhere. The mining and agricultural sectors in California were always characterized by relatively high productivity and by close links to the urban/industrial sector. And the historical structure of the state's farm sector did not encourage the growth of a large "surplus population" in the California countryside.

state's labor force compared with 38 percent nationally. And as in the United States, the employment share of the primary sector declined over time in California. It is notable that although the absolute number of workers in extractive activities declined nationally after 1910, it continued to expand in the state until the 1930s. Thereafter, extractive employment in California fell for four decades before beginning to rebound in the 1970s.

In contrast to agriculture and manufacturing, California's distribution and service sectors have constituted a larger share of the labor force than in the country as a whole. In 1900, services were the largest sector in the state, making up 45 percent of the California labor force, whereas they accounted for about 35 percent of the national total at this date. By 1960, the sector's share had risen to 63 percent in California, and in the United States as a whole, it had climbed to just under 60 percent. Thus, California led the way in the transition to becoming a service economy, and all indications are that its service sector will continue to grow faster than its manufacturing sector or its labor force as a whole.

The historically larger size of California's service sector was due to several factors:

- The combination of the high income elasticity of demand for services with the state's historically high levels of per capita incomes;
- California's long-standing role, based on its earlier settlement, as a financial and trade center of the greater western region;
- Its attractions as a major tourist destination; and
- Its emergence over the 20th century as a global center of the entertainment industry.

The traditionally smaller role of manufacturing in California is best viewed not as a sign of economic weakness or "backwardness" but rather as the product of the strength of its primary and tertiary sectors.

Figure 10.1 charts the manufacturing sector's share in total earnings in California and the entire United States over the 1929–1998 period. As it shows, in 1929, the manufacturing sector accounted for less than 17 percent of earnings in California compared with over 25 percent

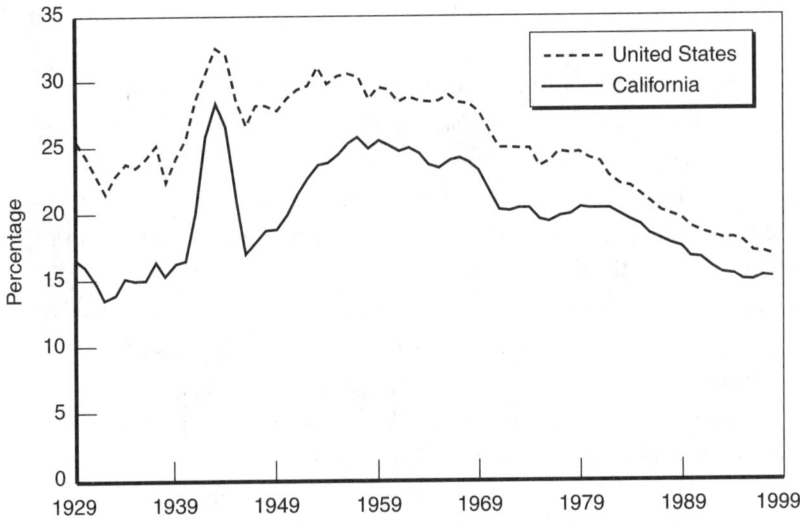

Figure 10.1—Manufacturing Sector's Share of Earnings in California and the United States, 1929–1998

nationally.[4] During the state's great boom between 1939 and 1958, the manufacturing sector was a driving force in California's economic growth. The most dramatic changes occurred during World War II, when military demands led earnings in California's manufacturing sector to soar. At the peak in 1943, this sector made up over 28 percent of total earnings in the state, a share never reached before the war and never surpassed thereafter. For the first time, manufacturing became the most important sector in the state. However, the earnings data also confirm that the role of the manufacturing sector in the economy has fallen since the mid-1950s. The relative contraction was gradual over the late 1950s and through most of the 1960s but accelerated at the end of that decade. By the mid-1970s, manufacturing had lost the title of California's most important sector. Manufacturing's share of total California earnings has gradually fallen, reaching levels close to those prevailing in the late 1930s.

[4]The data are from the BEA Regional Accounts Data, http://www.bea.doc.gov/bea/dr/spitbl-g.htm.

Data from the late 1990s show signs that the share reached a plateau of roughly 15 percent of the state's earnings.

One lesson from history is that California has long been less dependent on manufacturing than has the nation as a whole. Although the sector's share in the state has been relatively smaller, California experienced spectacular growth throughout the 20th century. A strong case can be made that manufacturing was the major driving force in this expansion only during the 1940s and 1950s. Since 1960, manufacturing's share in the state's aggregate economy has contracted, yet the California economy has continued to flourish. This pattern does not imply that state policymakers should actively encourage "deindustrialization" but rather that the California economy has not traditionally depended on manufacturing as its sole or primary "engine of growth." Nonetheless, its continued prosperity remains important if the state's economy is to continue to sustain its balanced growth path and to avoid becoming overly dependent on a single sector such as services to drive income and employment expansion.

11. Conclusions

The long-run growth of California's manufacturing sector has been nothing short of spectacular. In 1997, California manufacturing produced $204.1 billion of value added, over *one hundred times* (110) its real value in 1899 and employed about 1,193,000 production workers, 17 times the level of 1899. Put differently, this employment growth represents over one million more jobs than in the century before.

Over the past 150 years, California's population and its manufacturing sector have expanded together in what often appears as a mutually reinforcing, increasing-returns process. By most accounts, the small size of the region's population and markets in the late 19th and early 20th centuries limited local development of the emerging large-scale or mass production industries. They also constrained the scope and diversity of manufacturing activities carried out in the state. When population growth began to accelerate after 1900, local industrial expansion followed. Manufacturing played a key reinforcing role by creating jobs for many of the new Californians and reducing dependence on industrial imports.

During the mid-20th century, manufacturing took the lead, and its more rapid expansion made a major contribution to stimulating the state's vigorous population growth. This was the crucial period when California first stepped prominently on the world stage, becoming the most populous state in the early 1960s and the largest industrial state in the mid-1970s. Signs of the growing maturity of the state's manufacturing sector were evident not only in its vastly increased scale but also in its greater scope and diversity. Whereas at the beginning of the 20th century, large segments of the manufacturing sector were missing in the state, virtually every four-digit industry was present by 1997. And no single industry accounted for more than 5–6 percent of total output.

In more recent decades, the growth of the industrial sector has slowed and many of the state's manufacturers have turned to outside markets, such as the military and foreign exports. Such efforts, together with the expansion of lower-wage manufacturing activities, helped stave off the onset of deindustrialization. This process, which began to seriously affect the national economy in the 1970s and 1980s, was not fully felt in California until the downturn in early 1990s. Today, the role of the manufacturing sector in the state's growth is no longer that of the "leading man," as it was in the 1940s, 1950s, and 1960s. It is perhaps now better characterized as that of a major supporting player or key member of an ensemble cast.

In many ways, the evolution of the manufacturing sector has mirrored and continues to mirror the growth of the California economy as a whole. Manufacturing experienced the overall economy's shift away from a natural-resource base toward a knowledge or high-technology base. It has become an incubator for new innovative industries and startup firms just as California has become the birthplace for many new economic, political, and social trends. The manufacturing sector's productivity and labor returns have converged to the national averages just as the state's per capita income has converged to the national figure. Its changing structure has also reflected the increasing polarization of the state's economy into separate sectors employing highly educated, highly paid workers and others employing less-skilled, less-well-paid workers. And many of the distinguishing features of the state's industrial development have provided models that have been copied nationally and, indeed, globally. These include the pioneering use of new power sources and pursuit of new markets. Moreover, the state's traditional emphasis on entrepreneurship, innovation, and a small-plant culture has also returned into vogue.

A greater understanding of the state's economic history helps correct the myopia that plagues much contemporary thinking. Without a long-run view, every piece of good news (e.g., the growing popularity of electronic commerce) looks like the invention of a totally "new economy" and every piece of bad news looks like an unprecedented crisis. A study of the past reveals that many of the problems of the recent decade, including military cutbacks, energy shortages, and episode of employment volatility, have occurred periodically throughout California's history. A long view also helps reveal which challenges and opportunities are truly new—such as the state's increasing reliance on foreign markets to sell its manufactured products and the "disappearing middle" between low-wage and high-technology manufacturing sectors. The changes may present challenges that the state has limited experience addressing. Finally, the historical evidence indicates that, contrary to many contemporary claims, there is little to suggest that the 1990s was a period of especially rapid structural change. Indeed, change has been the one true constant in California's remarkable industrial transformation over the past 150 years.

Bibliography

Atack, Jeremy, "Economies of Scale and Efficiency Gains in the Rise of the Factory in America, 1820–1900," in Peter Kilby, ed., *Quantity and Quiddity: Essays in U.S. Economic History*, Wesleyan University Press, Middleton, Connecticut, 1987, pp. 286–335.

Balke, Nathan S., and Robert J. Gordon, "The Estimation of Prewar Gross National Product: Methodology and New Evidence," *Journal of Political Economy*, Vol. 97, No. 1, 1989, pp. 38–92.

Cain, Louis P., and Donald G. Paterson, "Biased Technical Change, Scale, and Factor Substitution in American Industry, 1850–1919," *Journal of Economic History*, Vol. 46, March 1986, pp. 153–164.

California Department of Finance, *California Statistical Abstract*, Sacramento, California, various years.

California Division of Labor Statistics and Research, *Changing Patterns in California Manufacturing Employment*, San Francisco, California, August 1967.

California State Mining Bureau, *Petroleum Industry of California*, Bulletin 69, State Printing Office, Sacramento, California, 1914.

Chandler, Alfred, *The Visible Hand: The Managerial Revolution in American Business*, Belkap Press, Cambridge, Massachusetts, 1977.

Clayton, James L., "Defense Spending: Key to California's Growth," *Western Political Quarterly*, Vol. 15, 1962, pp. 280–293.

Clayton, James L., "The Impact of the Cold War on the Economies of California and Utah, 1946–1965," *Pacific Historical Review*, Vol. 36, 1967, pp. 449–473.

Cleland, Robert Glass, and Osgood Hardy, *March of Industry*, Powell Publishing Co., Los Angeles, California, 1929.

Delehanty, George E., *Nonproduction Workers in U.S. Manufacturing*, North-Holland, Amsterdam, 1968.

Devine, Warren, "From Shafts to Wires: Historical Perspective on Electrification," *Journal of Economic History*, Vol. 43, 1983, pp. 347–372.

Dodd, Donald B., *Historical Statistics of the States of the United States: Two Centuries of the Census, 1790–1990*, Greenwood Press, Westport, Connecticut, 1993.

Gordon, Margaret, *Employment Expansion and Population Growth*, University of California Press, Berkeley, California, 1954.

Granovetter, Mark, "Small Is Bountiful: Labor Markets and Establishment Size," *American Sociological Review*, Vol. 49, No. 3, June 1984, pp. 323–334.

Grether, E. T., et al., *The Steel and Steel-Using Industries of California*, California Reconstruction and Re-employment Commission, State Printing Office, Sacramento, California, 1946.

Hoffman, W. G., *The Growth of Industrial Economies*, Manchester University Press, Manchester, 1958.

Johnson, Mark, "The Political Economics of Labor Market Segmentation: The Development of the California Labor Market, 1850–1900," unpublished Ph.D. Dissertation in Economics, Stanford University, 1985.

Kidner, Frank, *California Business Cycles*, University of California Press, Berkeley, California, 1946.

Markusen, Ann, et al., *The Rise of the Gunbelt: The Military Remapping of Industrial America,* Oxford University Press, New York, 1991.

Matthews, Richard A., *Fordism, Flexibility, and Regional Productivity Growth*, Garland, New York, 1996.

McWilliams, Carey, *California: The Great Exception,* Berkeley, University of California Press, 1949, republished with a new introduction by Lewis H. Lapham, 1998.

Nash, Gerald R., "Stages of California's Economic Growth," *California Historical Review*, Vol. 51, No. 4, Winter 1972, pp. 315–330.

Niemi, Albert W., Jr., *State and Regional Patterns in American Manufacturing, 1860–1900*, Greenwood Press, Westport, Connecticut, 1974.

Niklason, C. R., *Commercial Survey of the Pacific Southwest,* U.S. Bureau of Foreign and Domestic Commerce, Domestic Commerce Series No. 37, GPO, Washington, D.C., 1930.

Olmstead, Alan L., and Paul W. Rhode, "An Overview of California's Agricultural Mechanization, 1870–1930," *Agricultural History*, Vol. 62, No. 3, 1988, pp. 86–112.

Porat, Marc Uri, *The Information Economy: Sources and Methods for Measuring the Primary Information Sector*, OT Special Publication 77-12 (1), U.S. Department of Commerce, Office of Telecommunications, Washington, D.C., May 1977.

Saxenian, AnnaLee, *Regional Advantage: Culture and Competition in Silicon Valley and Route 128,* Harvard University Press, Cambridge, Massachusetts, 1994.

Shapira, Philip Paul, *Industry and Jobs in Transition: A Study of Industrial Restructuring Displacement in California*, Ph.D. Dissertation in City and Regional Planning, University of California, Berkeley, December 1986.

Thompson, Warren S., *Growth and Changes in California's Population,* Haynes Foundation, Los Angeles, California, 1955.

U.S. Bureau of the Census, *1947 Census of Manufactures*, Vol. I, *General Summary*, Vol. II, *Statistics by Industry*, Vol. III, *Statistics by States*, GPO, Washington, D.C., 1949–1950.

U.S. Bureau of the Census, *1954 Census of Manufactures*, Vol. I, *Summary Statistics*, Vol. II, *Statistics by Industry*, 2 Pts.; Vol. III, *Area Statistics*, GPO, Washington, D.C., 1957.

U.S. Bureau of the Census, *1958 Census of Manufactures*, Vol. I, *Summary and Subject Statistics;* Vol. II, *Industry Statistics*, 2 Pts.; Vol. III, *Area Statistics*, GPO, Washington, D.C., 1961.

U.S. Bureau of the Census, *1963 Census of Manufactures*, Vol. I, *Summary and Subject Statistics;* Vol. II, *Industry Statistics*, 2 Pts.; Vol. III, *Area Statistics*, GPO, Washington, D.C., 1965–1966.

U.S. Bureau of the Census, *1967 Census of Manufactures,* Vol. I, *Summary and Subject Statistics;* Vol. II, *Industry Statistics*, Pt. 1, *Major Groups, 20–24;* Pt. 2, *Major Groups, 25–33;* Pt. 3, *Major Groups, 34–39 and 19;* Vol. III, *Area Statistics*, Pt. 1, *Alabama-Montana;* Pt. 2, *Nebraska-Wyoming*, GPO, Washington, D.C., 1966.

U.S. Bureau of the Census, *1972 Census of Manufactures*, Vol. I, *Subject and Special Statistics;* Vol. II, *Industry Statistics*, 3 Pts., Vol. III, *Geographic Area Statistics*, Pt. 1, *Alabama-Montana;* Pt. 2, *Nebraska-Wyoming*, GPO, Washington, D.C., 1976.

U.S. Bureau of the Census, *1977 Census of Manufactures*, Vol. I, *Subject Statistics;* Vol. II, *Industry Statistics,* 2 Pts., Vol. III, *Geographic Area Statistics*, GPO, Washington, D.C., 1981.

U.S. Bureau of the Census, *1980 Census of Population*. Vol. I, *Characteristics of the Population*, Ch. D, *Detailed Population Characteristics*, Pt. 6, *California,* PC 80-1-D6, GPO, Washington, D.C., 1983.

U.S. Bureau of the Census, *1980 Census of Population*, Vol. I, *Characteristics of the Population*, Ch. D., *Detailed Population Characteristics*, Pt. 1, *United States Summary,* PC 80-1, GPO, Washington, D.C., 1984.

U.S. Bureau of the Census, *1982 Census of Manufactures*, Vol. I, *Subject Statistics*; Vol. II, *Geographic Area Statistics;* Vol. II, *Industry Statistics*, Pt. 1, *Major Group 20–26;* Pt. 2, *Major Group 27–34;* Pt. 3, *Major*

Group 35–39; Vol. III, *Geographic Area Series*, Pt. 5, *California*, GPO, Washington, D.C., 1989–1990.

U.S. Bureau of the Census, *1983 Annual Survey of Manufactures: Origin of Exports of Manufactured Products*, M83 (AS)-5, Washington, D.C., March 1986.

U.S. Bureau of the Census, *1984 Annual Survey of Manufactures: Statistics for Industries and Industry Groups,* M84 (AS)-1, Washington, D.C., July 1986.

U.S. Bureau of the Census, *1987 Census of Manufactures*, Vols. 1–7, *Subject Series*; Vol. 20A-39D, *Industry Series*; *Geographic Area Series*, Vol. 5, *California*, GPO, Washington, D.C., 1990–1992.

U.S. Bureau of the Census, *1990 Census of Population, Social and Economic Characteristics: California,* 1990 CP-2-6, GPO, Washington, D.C., 1993a.

U.S. Bureau of the Census, *1990 Census of Population, Social and Economic Characteristics: United States,* 1990 CP-2-1, GPO, Washington, D.C., 1993b.

U.S. Bureau of the Census, *1992 Census of Manufactures. General Summary*; Vol. 20A–39D, *Industry Series; Geographic Area Series*, Pt. 5, *California*, GPO, Washington, D.C., 1996.

U.S. Bureau of the Census, *1997 Census of Manufactures. General Summary; Geographic Area Series: California*, GPO, Washington, D.C., 2000–2001.

U.S. Bureau of the Census, *1997 Economic Census Data,* available at http://www.census.gov/epcd/www/econ97.html.

U.S. Bureau of the Census, *Annual Survey of Manufactures*, GPO, Washington, D.C., various years, 1949–1999.

U.S. Bureau of the Census, *Biennial Census of Manufactures 1921*, GPO, Washington, D.C., 1924.

U.S. Bureau of the Census, *Biennial Census of Manufactures 1923*, GPO, Washington, D.C., 1926.

U.S. Bureau of the Census, *Biennial Census of Manufactures 1925*, GPO, Washington, D.C., 1928.

U.S. Bureau of the Census, *Biennial Census of Manufactures 1927*, GPO, Washington, D.C., 1930.

U.S. Bureau of the Census, *Biennial Census of Manufactures 1931*, GPO, Washington, D.C., 1935.

U.S. Bureau of the Census, *Biennial Census of Manufactures 1933*, GPO, Washington, D.C., 1936.

U.S. Bureau of the Census, *Biennial Census of Manufactures 1935* (and 3 Supplements), GPO, Washington, D.C., 1937.

U.S. Bureau of the Census, *Biennial Census of Manufactures 1937*, 2 Pts., GPO, Washington, D.C., 1939.

U.S. Bureau of the Census, *Census of Electricity: Central Electric Light and Power, 1917*, GPO, Washington, D.C., 1920.

U.S. Bureau of the Census, *Census of Electricity: Central Electric Light and Power, 1922*, GPO, Washington, D.C., 1925.

U.S. Bureau of the Census, *Census of Electricity: Central Electric Light and Power, 1927*, GPO, Washington, D.C., 1930.

U.S. Bureau of the Census, *Census of Electricity: Central Electric Light and Power, 1932*, GPO, Washington, D.C., 1935.

U.S. Bureau of the Census, *Census of Electricity: Central Electric Light and Power, 1937*, GPO, Washington, D.C., 1940.

U.S. Bureau of the Census, *Census of Manufactures, 1905*. Pt. I, *United States by Industries*; Pt. II, *States and Territories*, GPO, Washington, D.C., 1907.

U.S. Bureau of the Census, *Census of Manufactures, 1914. Abstract;* Vol. I, *Reports by States;* Vol. II, *Reports for Selected Industries* GPO, Washington, D.C., 1918–1919.

U.S. Bureau of the Census, *Census of Manufactures: 1937, Man-Hour Statistics for 105 Selected Industries*, Washington, D.C., December 1939.

U.S. Bureau of the Census, *Census of Manufactures: 1939, Man-Hour Statistics for 171 Selected Industries*, Washington, D.C., December 1941.

U.S. Bureau of the Census, *Exports from Manufacturing Establishments: 1987,* ASM Analytical Report Series, AR87-1, Washington, D.C., February 1991.

U.S. Bureau of the Census, *Exports from Manufacturing Establishments: 1988 and 1989*, ASM Analytical Report Series, AR89-1, Washington, D.C., November 1992.

U.S. Bureau of the Census, *Exports from Manufacturing Establishments: 1990 and 1991*, ASM Analytical Report Series, AR91-1, Washington, D.C., December 1994.

U.S. Bureau of the Census, *Exports from Manufacturing Establishments: 1997*, AR(97)-1, December 2000, available at http://www.census.gov/mcd/ar97.pdf.

U.S. Bureau of the Census, *Fifteenth Census of the United States, Manufactures, 1929*, Vol. I, *General Report*, Vol. II, *Reports by Industry*, Vol. III, *Reports by States*, GPO, Washington, D.C., 1932.

U.S. Bureau of the Census, *Fourteenth Census of the United States, Manufactures, 1919*, Vol. VIII, *General Report*, Vol. IX, *Reports by States*, Vol. X, *Reports for Selected Industries*, GPO, Washington, D.C., 1921–1923.

U.S. Bureau of the Census, *Historical Statistics of the United States: Colonial Times to 1970*, 2 Vols., GPO, Washington, D.C., 1975.

U.S. Bureau of the Census, "Shipments of Defense-Oriented Industries," *Current Industrial Reports,* MA-175, Washington, D.C., annually, 1965–1977.

U.S. Bureau of the Census, "Shipments to Federal Government Agencies," *Current Industrial Reports,* MA-175, Washington, D.C., annually, 1978–1983.

U.S. Bureau of the Census, *Sixteenth Census of the United States, 1940, Manufactures, 1939,* Vol. I, *General Reports*, Vol. II, *Reports by Industry*, 2 Pts.; Vol. III, *States and Outlying Areas,* GPO, Washington, D.C., 1941–1942.

U.S. Bureau of the Census, "Survey of the Origin of Exports by Manufacturing Establishments, 1969," *Current Industrial Reports,* M161 (69)-2, Washington, D.C., January 7, 1971.

U.S. Bureau of the Census, "Survey of the Origin of Exports by Manufacturing Establishments in 1971," *Current Industrial Reports,* M161 (71)-2, Washington, D.C., March 1974.

U.S. Bureau of the Census, "Survey of the Origin of Exports by Manufacturing Establishments in 1972," *Current Industrial Reports,* M161 (72)-2, Washington, D.C., November 1974.

U.S. Bureau of the Census, "Survey of the Origin of Exports of Manufactured Products: 1960," *Current Industrial Reports*, M161 (60)-1, Washington, D.C., May 2, 1962.

U.S. Bureau of the Census, "Survey of the Origin of Exports of Manufactured Products: 1963," *Current Industrial Reports*, M161 (63)-1, Washington, D.C., June 2, 1965.

U.S. Bureau of the Census, "Survey of the Origin of Exports of Manufactured Products: 1966," *Current Industrial Reports,* M161 (66)-1 (Rev), Washington, D.C., January 17, 1968.

U.S. Bureau of the Census, *Thirteenth Census of the United States, Manufactures, 1909*, Vol. 8, *General Report*, Vol. 9, *Reports by States;*

Vol. 10, *Reports for Principal Industries*, GPO, Washington, D.C., 1912.

U.S. Bureau of Economic Analysis home page, http://www.bea.doc.gov/bea/dr/spitbl-g.htm.

U.S. Bureau of Labor Statistics home page, http://stats.bls.gov/datahome.htm.

U.S. Census Office, *Statistical View of the United States: A Compendium of the Seventh Census,* GPO, Washington, D.C., 1854.

U.S. Census Office, *Eighth Census of the United States, 1860,* Vol. 3, *Manufactures,* GPO, Washington, D.C., 1865.

U.S. Census Office, *Ninth Census of the United States, 1870,* Vol. 3, *Wealth and Industry,* GPO, Washington, D.C., 1872.

U.S. Census Office, *Tenth Census of the United States, 1880,* Vol. 2, *Manufacturing of the United States,* GPO, Washington, D.C., 1883.

U.S. Census Office, *Eleventh Census of the United States, 1890,* Vol. 11, *Manufacturing Industries,* Pt. I, *Totals for States and Industries;* Vol. 12, *Manufacturing Industries,* Pt. II, *Statistics for Cities;* Vol. 13, *Manufacturing Industries,* Pt. III, *Selected Industries*, GPO, Washington, D.C., 1895.

U.S. Census Office. *Twelfth Census of the United States*, 1900, Vol. 7, *Manufacturing*, Pt. 1 *United States by Industry*; Vol. 8, *Manufacturing*, Pt. 2 *States and Territories;* Vol. 9, *Manufacturing*, Pt. 3 *Special Reports on Selected Industries;* Vol. 10, *Manufacturing*, Pt. 4 *Special Reports on Selected Industries, Continued*, GPO, Washington, D.C., GPO, 1902.

U.S. Department of Commerce, *Regional Employment by Industry, 1940–1970*, GPO, Washington, D.C., 1975.

U.S. Interstate Commerce Commission, *Railways in the United States*, Pt. II, *A Forty Year Review of Changes in Freight Rates*, GPO, Washington, D.C., 1903.

U.S. Interstate Commerce Commission, *Statistics of Railways in the United States*, GPO, Washington, D.C., various years, 1887–1960.

Ullman, Edward L., "Regional Development and the Geography of Concentration," *Papers and Proceedings of the Regional Science Association,* 1958, pp. 179–1998.

Wheeler, Benjamin Ide, "A Forecast for California and the Pacific Coast," *Outlook*, Vol. 99, No. 23, September 1911, pp. 167–1974.

Williams, James C., *Energy and the Making of Modern California*, University of Akron Press, Akron, Ohio, 1997.

About the Author

PAUL W. RHODE

Paul W. Rhode is a professor of economics at the University of North Carolina, Chapel Hill, and a former visiting fellow at PPIC. His research focuses on the economic development of the American West over the 20th century. He has a B.A. from the University of California, Davis, and a Ph.D. in economics from Stanford University.

Other Related PPIC Publications

Rethinking the California Business Climate
Michael Dardia, Sherman Luk

California's Vested Interest in U.S. Trade Liberalization Initiatives
Jon D. Haveman

California's Rising Income Inequality: Causes and Concerns
Deborah Reed

Silicon Valley's New Immigrant Entrepreneurs
AnnaLee Saxenian

PPIC publications may be ordered by phone or from our website
(800) 232-5343 [mainland U.S.]
(415) 291-4400 [Canada, Hawaii, overseas]
www.ppic.org